LEARNING ECONOMICS

a practical workbook

LEARNING ECONOMICS

a practical workbook

Abhay B. Ghiara

DeVry Institute of Technology
DuPage Campus

Learning Economics: A Practical Workbook by Abhay Ghiara

ISBN: 0-201-32761-9

12345678910-CRS-0100999897

Copyright Acknowledgements

Grateful acknowledgement is made to the following sources for permission to reprint material copyrighted or controlled by them:

"Best of both worlds," an interview with Milton Friedman conducted by Brian Doherty, reprinted from *Reason*, v27, Jun 1995. Page: 32. Length: 7 page(s). Number: 9505162059. Publisher: Reason Foundation. ISSN: 0048-6906

"Unique perspective," an interview with John Kenneth Galbraith conducted by Shirin Sinnar and Valerie Jaffee, reprinted from *Harvard International Review,* v18 n2, Spring 1996. Page: 50. Length: 4 page(s). Number: 9604293493. Publisher: Harvard International Review. ISSN: 0739-1854

"Drift, diversion, and snake oil," an interview with Paul Krugman conducted by A.J. Vogl, reprinted from *Across the Board*, v31 n5, May 1994. Page: 25. Length: 6 page(s). Number: 9411170864. Publisher: Conference Board Inc.. ISSN: 1046-090X

"Are we really that lazy?" an interview with Juliet B. Schor conducted by Annetta Miller, reprinted from *Newsweek*, v119, Feb 17, 1992. Page: 42. Length: 2 page(s). Number: 9202172597. Publisher: Newsweek. ISSN: 0028-9604

"Building a better economy," an interview with Lester Thurow conducted by Sandra Hackman, reprinted from *Technology Review*, v99 n4, May-Jun 1996. Page: 41. Length: 6 page(s). Number: 9605021495. Publisher: Technology Review. ISSN: 0040-1692

About the Book

Learning Economics: A Practical Workbook is designed to complement any serious one-term economics textbook. It is organized in four sections:

- **Section One** contains topic by topic *problem-solving, graphical, and analytical exercises*. Economics can be best learned using a *learn-by-doing* approach. These exercises are layered to meet the needs of a wide variety of users. Instructors wishing to emphasize the core principles and their applications can choose from over forty exercises to assign to students. Other instructors, who incorporate the rudiments of algebra in their courses can also assign the exercises marked with the symbol ***MATH!*** Instructors can also incorporate the use of the Internet in their courses by assigning exercises marked with the symbol ***INTERNET!***

- **Section Two** contains rudimentary mathematical economics using simple linear relationships. My experience teaching students from engineering and technology-based educational backgrounds has taught me that the use of basic math in economics can help bring the subject into a clearer focus. Many students are pleasantly surprised to find that the seemingly dry concepts learned in their math courses have lively and practical applications in economics. This section brings students up to the level of mathematical sophistication required to answer the specially marked ***MATH!*** exercises.

- **Section Three** contains the most common economic statistics such as the GDP and the CPI that are taught in a typical one-term economics course. This section briefly describes how the student can use each economic statistic, presents graphs of data for the US economy from 1960 to the present, and provides *INTERNET* addresses as well as *HOTLINE* phone numbers where students can find both historical data as well as the very latest information just as it is released.

- **Section Four** contains recent interviews with leading economists of our time. Featured here, among others, are two of the greatest economists of our time: Milton Friedman, a believer in a free-market, minimal-government society, and John Kenneth Galbraith, a firm believer in a prominent role of the government in a capitalist society. The interviews serve to add a much needed human dimension to the introductory economics course. After reading these interviews, students begin to appreciate both the unity as well as the diversity in prominent economists' approaches to understanding social phenomena.

About the Author

Abhay B. Ghiara was trained at Northwestern University, where he studied under such notable economists as George Dalton and Robert Eisner. His experience teaching nontraditional students at Northwestern University's Continuing Education Program made him an active advocate and promoter of student-centered learning in economics. He is now Associate Professor of Economics at the DeVry Institute of Technology where he practices what he preaches.

CONTENTS

SECTION ONE — Exercises

SECTION TWO Mathematical Economics

SECTION THREE Economic Statistics

SECTION FOUR Interviews with Economists

SECTION ONE

Exercises

Name________________________________

Exercise 1.1

Consider the following alternative production possibilities:

1) Graph a production possibilities curve from the information given in the table

	Consumption goods	Investment goods
A	0	300
B	20	280
C	40	240
D	60	180
E	80	100
F	100	0

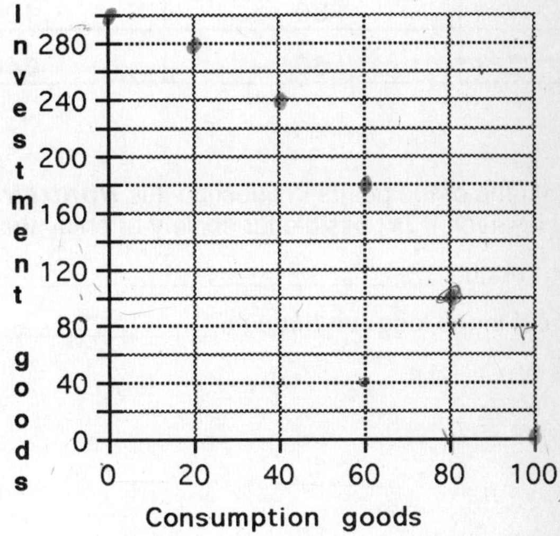

2) Fill in the blanks in the table below:

Units of Consumption goods	Opportunity Cost of producing Consumption goods
First 20 (point A to B)	20 Investment goods
Second 20 (point B to C)	40 Investment goods
Third 20 (point C to D)	60 Investment goods
Fourth 20 (point D to E)	80 Investment goods
Fifth 20 (point E to F)	100 Investment goods

3) What is the name of the general principle that the table above illustrates?

Law of increasing cost

4) Indicate whether the following points illustrate an underemployment of resources, full employment of resources or are unattainable:

	Consumption goods	Investment goods	Underemployment/ Full employment/ Unattainable
X	40	60	
Y	80	100	
Z	60	240	

5) One of the points in question 4 is ***unattainable***.
However, it ***is*** possible for society to enjoy that level of consumption in the short-run through __________________________________,

and in the long-run through __________________________________.

Name______________________________

Exercise 1.2

1) Graph the following

X	Y
1	4
2	9
3	14
4	19

2) Graph the following

X **# of students on campus**	Y **# of pizzas sold**
300	60
500	100
700	140
900	180

3) Graph the following

X **Quantity demanded per month**	Y **Price per compact disc**
10,000	$5
8,000	$8
6,000	$11
4,000	$14

Name________________________________

Exercise 1.3 ***MATH!***

1) Graph the following linear relationship and calculate the slope:

X	Y
4	5
7	10
10	15
13	20

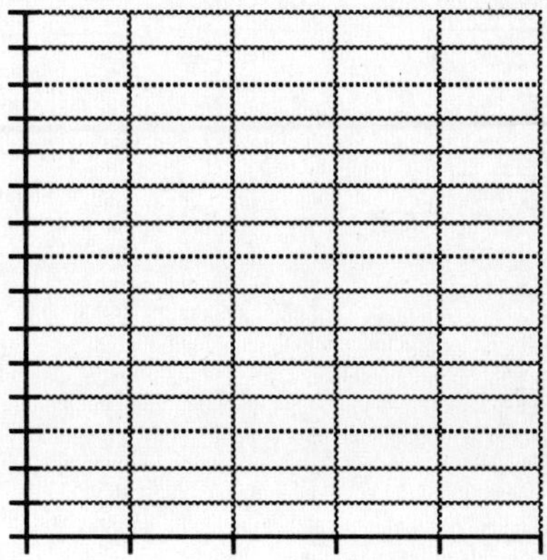

2) Graph the following linear relationship and calculate the slope:

X	Y
300	50
500	90
700	130
900	170

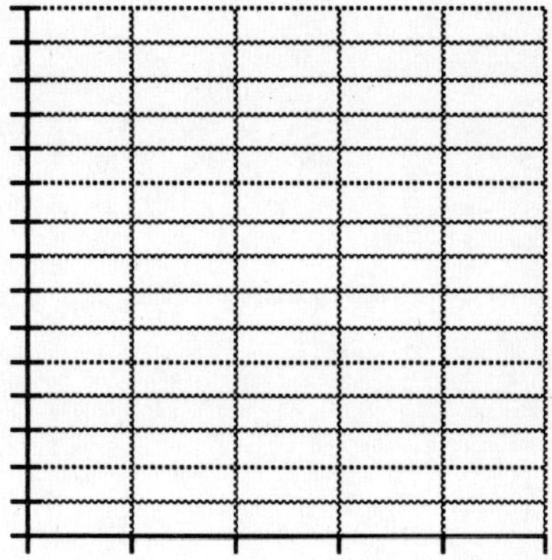

3) Graph the following linear relationship and calculate the slope:

X	Y
12,000	$7
10,000	$10
8,000	$13
6,000	$16

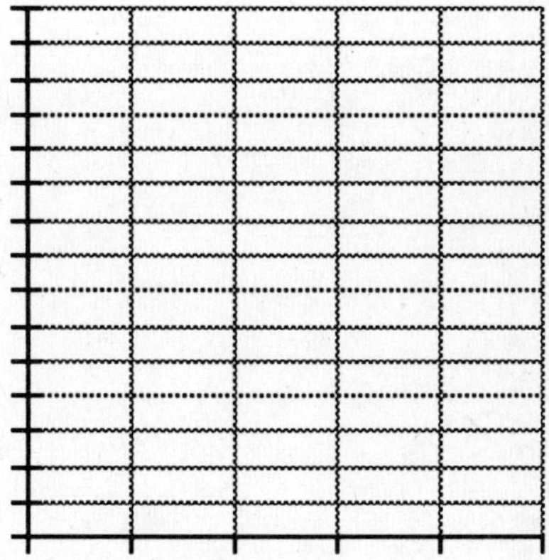

Name________________________________

Exercise 2.0

1) Which economic system involves the following characteristics? Circle the correct answer for each.

Private property and the protection of property rights	Pure Capitalism	Pure Command Socialism
Economic planning	Pure Capitalism	Pure Command Socialism
Competition	Pure Capitalism	Pure Command Socialism
Allocation by directives	Pure Capitalism	Pure Command Socialism
Centralized decision making	Pure Capitalism	Pure Command Socialism
Signaling, motivating and rationing roles of prices	Pure Capitalism	Pure Command Socialism
Limited government intervention	Pure Capitalism	Pure Command Socialism
Public ownership	Pure Capitalism	Pure Command Socialism
Self-interest	Pure Capitalism	Pure Command Socialism

2) Draw a diagram of the circular-flow model below. Be careful to label all components clearly and correctly:

3) Draw a diagram of the command pyramid model below. Be careful to label all components clearly and correctly:

Name______________________________

Exercise 3.1

The following demand and supply schedules refer to the market for VCRs at Jane and Tom's family-owned electronics store.

Price	Quantity Demanded	Quantity Supplied
$248	15	23
$239	20	20
$230	25	17
$221	30	14

1) Graph the demand and supply curves below on one graph:

2) What are the equilibrium price and equilibrium quantity?

239, 20

3) At price = $221, is there a shortage or surplus? What is its size?

4) At price = $248, is there a shortage or surplus? What is its size?

Name______________________________

Exercise 3.2

1) Consider the **market for wool sweaters**. For each example circle the correct answer.

New studies show that wool produces a number of mysterious allergies	increase in demand	decrease in demand	no change in demand
This year most working people get a big bonus and a raise	increase in demand	decrease in demand	no change in demand
Income taxes are increased	increase in demand	decrease in demand	no change in demand
Price of synthetic sweaters decreases	increase in demand	decrease in demand	no change in demand
Cost of dry cleaning and woolite increases	increase in demand	decrease in demand	no change in demand
The number of people living in the colder states increases	increase in demand	decrease in demand	no change in demand
Price of wool sweaters increases	increase in demand	decrease in demand	no change in demand
People expect the coming winter to be a particularly cold winter	increase in demand	decrease in demand	no change in demand

2) Draw a graph of the **market for wool sweaters** and clearly show an increase in demand and its effects on equilibrium price and quantity.

Name______________________________

Exercise 3.3

1) Draw a graph for the market for refrigerators below:

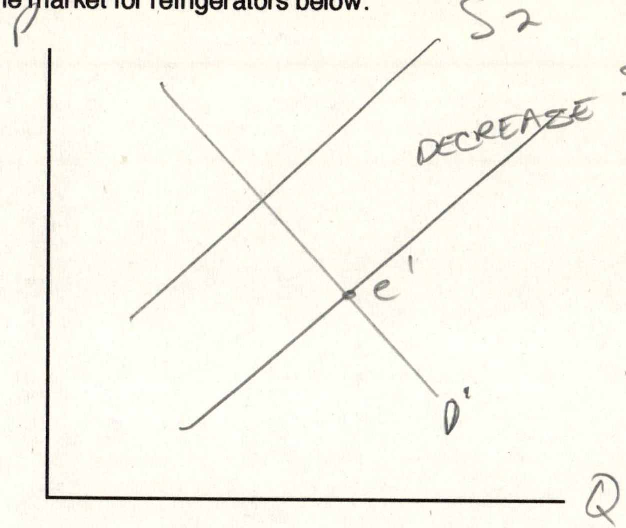

2) On the graph above, indicate what would happen if the cost of compressors (a component of refrigerators) were to increase.

increase in supply price will raise

3) What happens to equilibrium price?

INCREASES

4) What happens to equilibrium quantity?

DECREASES

5) Consider the market for **leather dress shoes**. For each example, circle the correct answer

Cost of leather used to manufacture dress shoes increases	increase in supply	decrease in supply	no change in supply
New technology makes it cheaper to manufacture leather dress shoes	increase in supply	decrease in supply	no change in supply
Price of leather dress shoes increases	increase in supply	decrease in supply	no change in supply
Several new producers of leather dress shoes start producing them	increase in supply	decrease in supply	no change in supply

Name________________________________

Exercise 3.4

In each of the graphs below, show a **shift** in **either** the **demand curve *or*** the **supply curve** as a result of the event presented. Label the new supply or demand curve clearly and show the direction of the change.

a) Health reports indicate that consuming fish rather than red meat prolongs life and prevents heart disease.

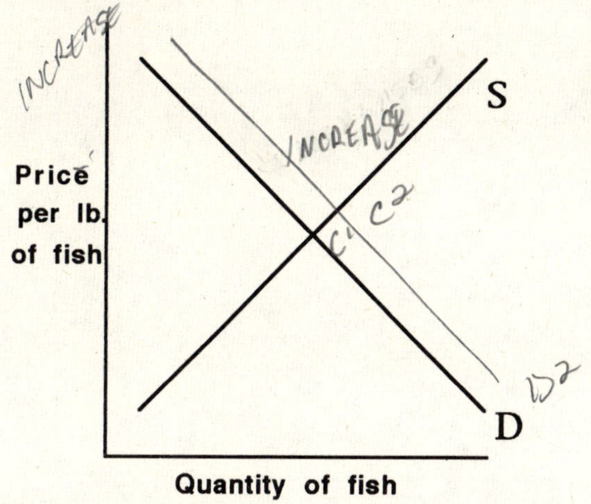

b) Technological improvement reduces the production costs of computers.

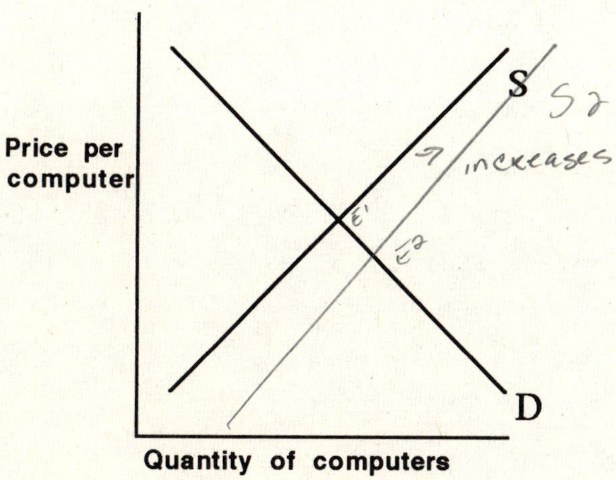

c) The price of aluminum used to manufacture airplanes increases.

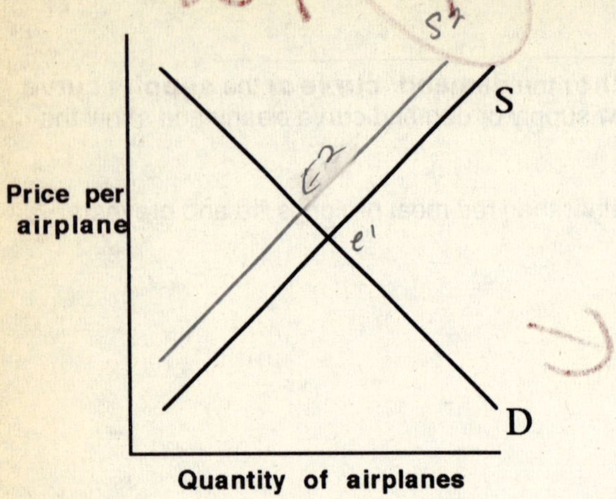

d) The price of hot dogs increases.

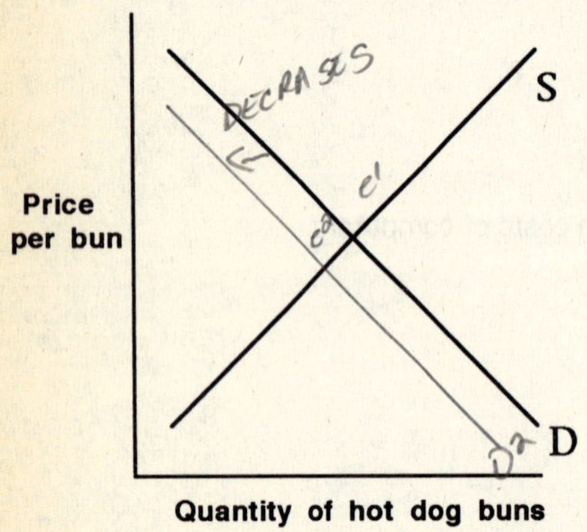

Name________________________________

Exercise 3.5 ***MATH!***

Consider the market for compact discs. Suppose the following are the demand and supply functions for compact discs:

$$Q_d = 150 - 5P$$
$$Q_s = 17 + 2P$$

where:

Q_d = quantity demanded
Q_s = quantity supplied
P = price per CD

1) Calculate the equilibrium price

2) Calculate the equilibrium quantity

3) What would happen if compact disc sellers priced compact discs at $20 per CD?

Name________________________________

Exercise 4.1

The following are recent estimates of the price elasticity of demand for air travel and coffee:

Product/Service	Price Elasticity of Demand
Air Travel	2.4
Coffee	.3

1) Which of the two is relatively elastic, which is relatively inelastic?

2) Suppose the price of air travel as well as coffee were to increase and other things remained unchanged:

a) Would **total revenue** from air travel sales increase or decrease?

b) Would **total revenue** from coffee sales increase or decrease?

3) By what percentage would the price of air travel have to decrease to boost sales by 20%?

4) By what percentage would the price of coffee have to decrease to boost sales by 20%?

5) Which of the two products shows a greater responsiveness of quantity demanded to changes in price?

Name____

Exercise 4.2

The following table and graph represent the demand schedule and demand curve for a particular product:

Point	Price	Quantity Demanded
A	$5	0
B	$4	1.5
C	$3	3.0
D	$2	4.5
E	$1	6.0

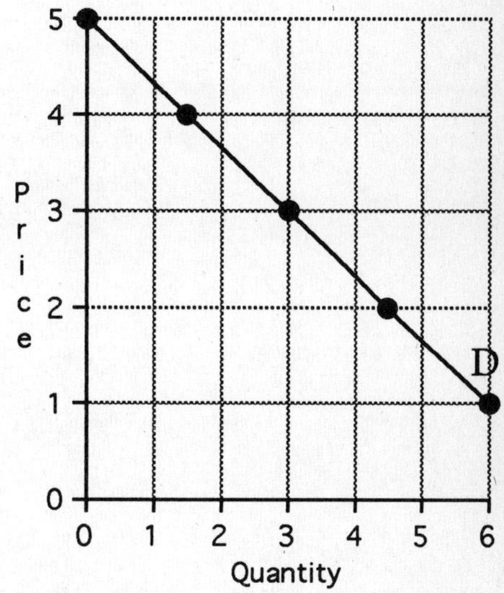

Fill in the cells in the tables below based on the demand schedule and / or the demand curve:

Point	Total Revenue
A	
B	
C	
D	
E	

Points	Coefficient of Demand Elasticity	Total Revenue is Increasing/ Decreasing/ Constant	Demand is: Elastic/ Inelastic/ Unitary Elastic
A to B			
B to C			
C to D			
D to E			

Name________________________________

Exercise 4.3

1) If the average price per compact disc decreased from $15.00 to $12.00 and the quantity demanded increased from 20,000 to 35,000, calculate the coefficient of demand elasticity.

2) If the coefficient of demand elasticity is 0.85 and the producer wants to raise total revenues by 20%, would the price have to be decreased or increased?

By what percent?

3) If the price of movie tickets is increased and total revenues increase along with it, that would mean that the demand for movies is:

4) Which would have a greater demand elasticity: a pair of Nikes or a pair of athletic shoes in general?

5) Would coffee filters have a elastic or inelastic demand? Why?

Name________________________________

Exercise 4.3

1) If the average price per compact disc decreased from $15.00 to $12.00 and the quantity demanded increased from 20,000 to 35,000, calculate the coefficient of demand elasticity.

2) If the coefficient of demand elasticity is 0.85 and the producer wants to raise total revenues, would the price have to be decreased or increased?

3) If the coefficient of demand elasticity is 0.85 and the producer wants to raise quantity demanded by 20%, how much (in percent) would the price have to be decreased?

4) If the price of movie tickets is increased and total revenues increase along with it, that would mean that the demand for movies is:

5) Which would have a greater demand elasticity: a pair of Nikes or a pair of athletic shoes in general?

6) Would coffee filters have a elastic or inelastic demand? Why?

Name________________________________

Exercise 4.4

1) If the average price of point-and-shoot cameras increased from $50 to $100, and the quantity supplied increased from 2,500 to 3,000, calculate the coefficient of supply elasticity.

2) If the coefficient of supply elasticity is 2.5, and the goal is to increase supply by 10%, by how much (as a percent) would the price have to be raised?

3) **The longer the time period, the greater is the elasticity of supply**. Is this statement true? If so, **why**?

Name________________________________

Exercise 5.1

1) Fill in the empty cells below and graph the TFC, TVC and TC curves.

Output	TFC	TVC	TC
0	$50	0	
1	$50	30	
2	$50	40	
3	$50	45	
4	$50	55	
5	$50	75	

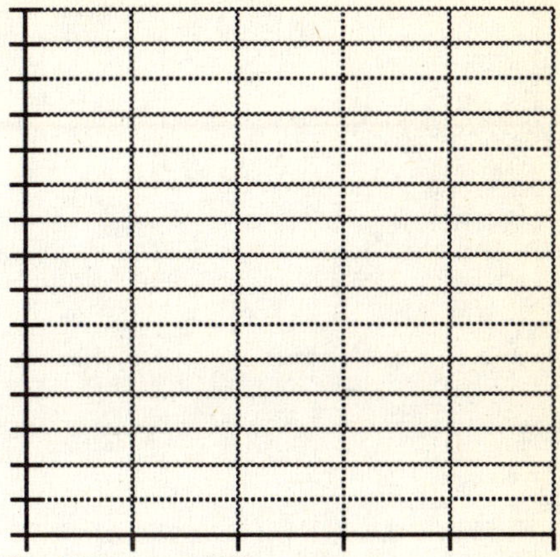

1) Fill in the empty cells below and graph the AFC, AVC, ATC, and MC curves.

Output	TFC	TVC	TC	AFC	AVC	ATC	MC
1	$50	40					
2	$50	50					
3	$50	55					
4	$50	65					
5	$50	85					
6	$50	130					

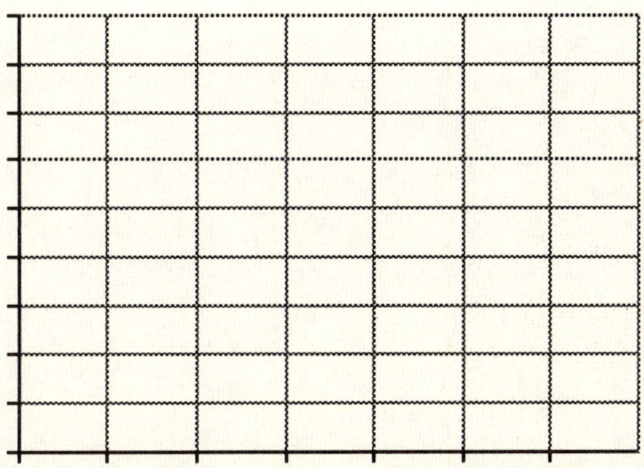

Name________________________________

Exercise 5.2

The following questions are based on the graph below which represents the cost curves for a firm in the short-run:

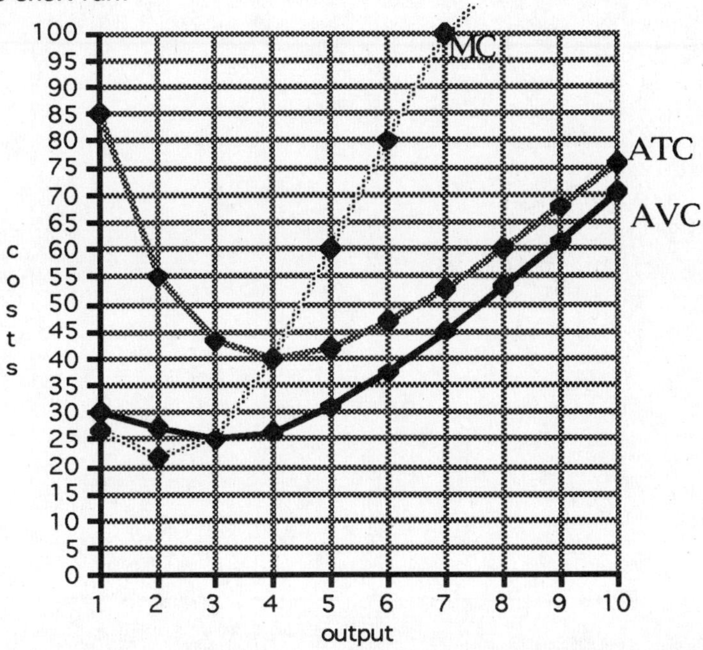

1) What is the price below which the firm will minimize its loss by shutting down in the short-run?

2) What is the price at which the firm will earn only a normal profit or zero economic profit?

3) If the firm is earning only normal profit (zero economic profit), calculate the following:

Total Revenue	Total Cost	Total Profit

4) If price equals $80 under pure (perfect) competition, shade in the total (economic) profit on the graph above and clearly label the region as PROFIT or ECONOMIC PROFIT.

5) If price equals $35 under pure (perfect) competition, will the firm be better off shutting down or continuing to operate, in the short-run?

6) If price equals $35 under pure (perfect) competition, will the firm make an economic profit, a normal profit (zero economic profit), or an economic loss, in the short-run?

7) What will be the total (economic) profit in the long-run under pure (perfect) competition?

Name____
Exercise 6.0

The following questions refer to the graph below:

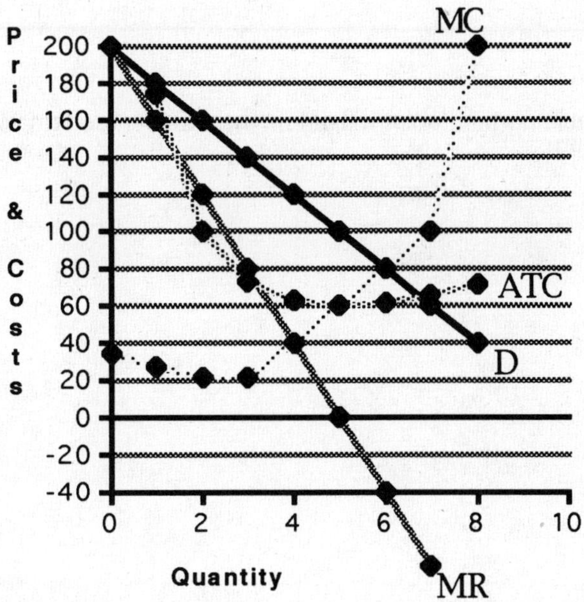

1) Does the graph above represent a firm under **pure competition**, or a firm with **market power**?

What aspect of the graph helped you decide that?

2) What is the profit-maximizing level of output here?

What rule did you use to figure this out?

3) Shade in the total (economic) profit on the graph above and clearly label the region as PROFIT or ECONOMIC PROFIT.

4) For an ***efficient allocation of resources***, what would the level of output have to be?

What rule did you use to figure this out?

5) In the **long run**, what important factor will determine whether the firm with market power can continue to earn its profit?

Name________________________________

Exercise 7.0

1) In the table below fill in the blank cells with **one** of the **four basic industry structures** that matches the characteristics listed:

Characteristics	**Industry Structure**
Many small sellers of differentiated products with no substantial barriers to entry	
One sole supplier of a unique product with substantial barriers to entry	
Many small sellers of identical products with no substantial barriers to entry	
Few large sellers of identical *or* differentiated products with substantial barriers to entry	

2) Calculate the **Nash Equilibrium** from the payoff matrix below:

Firm A's price strategies

Firm B's price strategies		$52.50	$70.00
	$52.50	Each firm earns a $70,000 profit	A earns $35,000 B earns $140,000
	$70.00	A earns $140,000 B earns $35,000	Each firm earns a $105,000 profit

3) Is there an incentive for **collusion** here?

Explain why and what can be accomplished by collusion in this case.

Name______________________________

Exercise 8.1

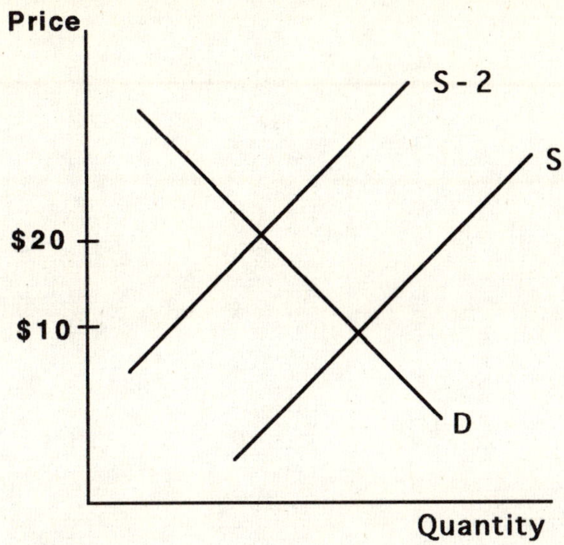

1) In the graph above, which supply curve would be more likely represent both private costs and external costs?

Why?

2) Which **one** out of the following three would you pick to solve the problem of pollution?

a) Pollution taxes
b) Emissions standards
c) Assigning property rights

Why did you pick this one over the others?

3) Think of an example of a public good and explain why you would classify it as a public good rather than classify it as a **pure private good** or a **private good with external benefits.**

Name______________________________________

Exercise 8.2 ***INTERNET!***

Visit the Economic Statistics Briefing Room of the White House at:

http://www.whitehouse.gov/fsbr/esbr.html

Click on Income

1) What is the poverty rate for the most recent period that information is available for?

2) How many people were classified as poor?

3) Is the poverty rate at present higher or lower than it was in 1964?

4) Are there more or less people living in poverty at present compared to the year 1964?

Name______________________

Exercise 9.1

1) Complete the table below:

	Counted as Unemployed?		If yes, which type of unemployment
Pauline, who is 15, has been actively looking for a job	Yes	[No]	
Mr. Johnston, a prison inmate, has not landed a job in the 4 years he has been in prison	Yes	[No]	
Jenny lost her job as a TV commercial writer when the economy moved into a recession	[Yes]	No	cyclical
Ms. Geneva's skills as an expert hand-made shoe maker were no longer enough to get her a good job	[Yes]	No	structural
Anthony quit his job as a sales representative and looked for a job that better suited his training as a neurosurgeon	[Yes]	No	FRICTIONAL
Patricia Van Doren was given a 20 hours a week, part-time job in place of the full-time job she had	Yes	[No]	
John Travolta decided to stay home, take care of his son, cook meals and not look for work	Yes	[No]	
Hubert H. Huffingham, Jr. looked for a job for years and finally gave up looking	Yes	[No]	

2) Given the table of information below, calculate the following:

number unemployed:	6500
number employed:	120,000

The labor force is 124,500

The unemployment rate is 5.14

Name________________________________

Exercise 9.2

The following table shows the actual numbers for the **consumer price index** published by the Bureau of Labor Statistics. The base period is the average for 1982 to 1984.

Year	CPI
1988	118.3
1989	124.0
1990	130.7
1991	136.2
1992	140.3
1993	144.5
1994	148.2
1995	152.4
1996	156.9

1) Calculate the inflation rate between the base period and 1996

2) Calculate the inflation rate between 1988 and 1989

3) Calculate the inflation rate between 1995 and 1996

4) If it is forecasted that prices will increase 3% from 1996 to 1997, what would the CPI in 1997 be?

Name________________________________

Exercise 9.3

1) Suppose the inflation rate = 3% a year for the next 5 years. Using year 1 as the base period, calculate the price index for each year:

Years	**Price Index**
Year 1	
Year 2	
Year 3	
Year 4	
Year 5	

2) Now suppose you earn a salary of $35,00 a year for the next five years. Calculate your real salary for each of the 5 years (using year 1 as the base period):

Year	**Nominal Salary**	**Real Salary**
Year 1	$35,000	
Year 2	$35,000	
Year 3	$35,000	
Year 4	$35,000	
Year 5	$35,000	

3) By what percent would your salary have to increase each year to keep your real salary at $35,000?

Name________________________________

Exercise 9.4

The table below shows

a) the actual average nominal wage per hour in the US

b) CPI for each year with base period = 1982-84

1) Calculate real wage and fill in the cells below:

Year	Nominal wage ($ per hour)	CPI	Real wage ($ per hour)
1960	2.43	29.6	
1970	4.25	38.8	
1980	9.70	82.4	
1990	16.30	130.7	

2) Complete the table below

Years	% change in nominal wage	% change in real wage
1960-1970		
1970-1980		
1980-1990		

3) You worked for Tight & Belt Co. from 1980 to 1990. From 1980 to 1989 you earned the same wage of $9.00 an hour. In 1990 you got a 40% raise in your hourly wage. Is your real wage in 1990 higher than your real wage in 1980? SHOW ALL YOUR WORK. (Use CPI information from the table.)

Name______________________________

Exercise 9.5

The table below contains data on output and prices in an economy that **only produces fast food!**

1996

	Price	Quantity
Hamburgers	$2.20	100
Shakes	$2.00	300
Fries	$0.95	200

1997

	Price	Quantity
Hamburgers	$2.80	100
Shakes	$2.30	300
Fries	$1.25	200

1) Calculate the Nominal GDP for 1996:

2) Calculate the Nominal GDP for 1997:

3) Calculate the percentage increase in Nominal GDP from 1996 to 1997:

4) Assuming 1996 is the base period, calculate the Real GDP for 1996:

5) Assuming 1996 is the gase period, calculate the Real GDP for 1997:

6) Calculate the percentage increase in Real GDP from 1996 to 1997:

7) Compare your answers to (3) and (6). Is Nominal GDP or Real GDP a more accurate representation of the growth of output from 1996 to 1997?

Name________________________

Exercise 9.6

1) From the data (in billions) given below calculate the value of the GROSS DOMESTIC PRODUCT (GDP):

Capital consumption allowances	800
Personal consumption expenditures	5000
Rental income	37
Net exports	-102
Net interest	400
Proprietors' income	512
Gross private domestic investment	1280
Corporate profits	567
Employee compensation	4532
Government purchases	1265
Indirect business taxes	595

$7443

2) Fill in the blanks in the table below: (This question is not related to question 1)

Year	Nominal GDP (billions of $)	Implicit Price Deflator	Real GDP (billions of $)
1971	1097	37	2964
1981	3035	79	3842
1991	5723	118	4850

3) Use the information from question 2 to complete the table below:

Year	% change in nominal GDP	% change in Implicit Price Deflator	% change in real GDP
1971 to 1981	176.79%	113.59%	29.69%
1981 to 1991	88.59%	49.49%	26.2%

Name________________________________

Exercise 9.7 ***INTERNET!***

Visit the Economic Statistics Briefing Room of the White House at:

http://www.whitehouse.gov/fsbr/esbr.html

Click on Employment, Unemployment and Earnings

1) What is the unemployment rate in the most recent period?

2) How many people are unemployed? (Are the figures in thousands, millions, billions....?)

3) Between which two years did we experience the highest unemployment rate during the last 10 years?

4) During which year did we have the lowest unemployment rate during the last 10 years?

5) In the most recent period, what was the most common reason for unemployment?

Click on Prices

6) What was the Inflation rate (as measured by the percent change in the Consumer Price Index) in the most recent period?

7) During which 2 years was the Inflation rate the highest during the last 10 years?

8) Would inflation be lower or higher in the most recent period if energy and food prices were excluded from the calculations?

9) What was the percent change in the Producer Price Index in the most recent period?

10) Did the Producer Price Index increase more or less the the Consumer Price Index?

Click on Output

11) How much (as a percent) did real GDP increase or decrease in the most recent quarter?

12) Name the most recent year and the quarter when real GDP decreased.

13) Name the years and quarters when real GDP decreased for at least two consecutive quarters.

Name____________________________________

Exercise 10.1

1) On graph (A) below, indicate what would happen if government spending increased.

On graph (B) below, indicate what would happen if businesses reduced their investment spending.

(A)

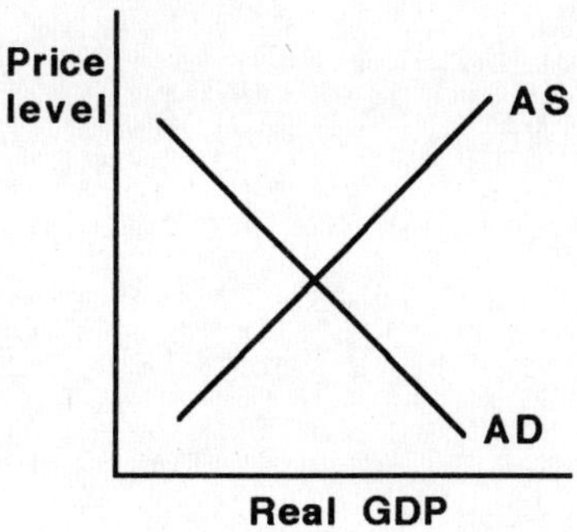

(B)

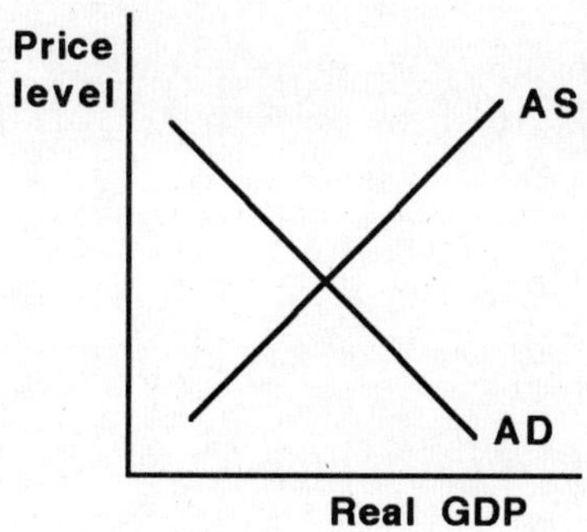

2) On graph (C) below, indicate what would happen if the price of oil suddenly increased.

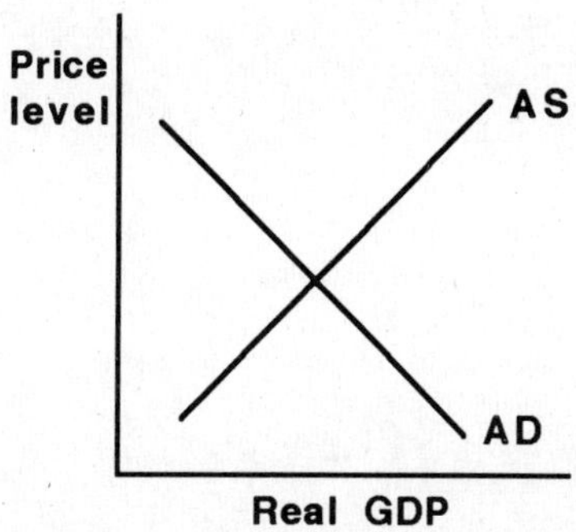

3) Fill in the table with the words, “increases” or “decreases” to show the effects of the shifting **A D** or **A S** curves on equilibrium price level and equilibrium real GDP:

	Price level	Real GDP
Graph (A)		
Graph (B)		
Graph (C)		

Name_______________________________________

Exercise 10.2

(A)

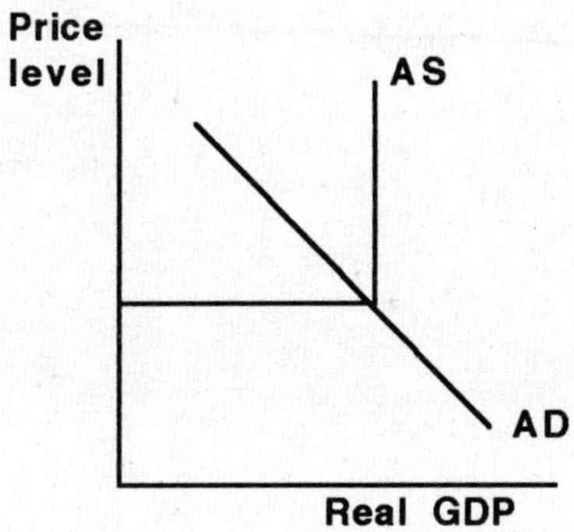

(B)

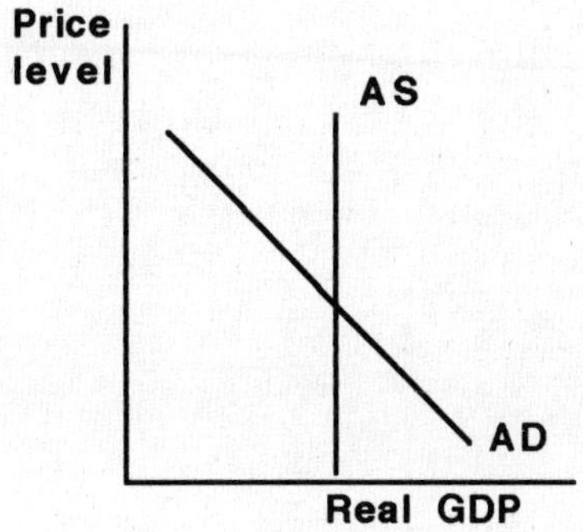

1) Which graph above represents the classical model and which one represents the Keynesian model?

WHY?

2) On both the graphs above, show what would happen if consumers spent less of their income.

3) In graph (A), what happens to prices and wages as a result of the changes you have made?

4) In graph (A), what happens to real GDP as a result of the changes you have made?

5) In graph (B), what happens to prices and wages as a result of the changes you have made?

6) In graph (B), what happens to real GDP as a result of the changes you have made?

Name________________________________

Exercise 11.1

1) Complete the table below (all figures are in billions of dollars):

Total Income and Output (GDP)	Planned Consumption	Planned Investment	Total Planned Expenditures	Tendency of Output
3400	3200	600	3800	increase
4400	4000	600	4600	increase
5400	4800	600	5400	no change
6400	5600	600	6200	decrease

2) Calculate the marginal propensity to consume (MPC)

$\frac{800}{1000}$ = mpc .8

3) Calculate the marginal propensity to save (MPS)

.8 − 1 = .2

4) At what level of total output is the economy at an equilibrium?

5400

5) Why is a level of total output = $3400 billion not an equilibrium? Discuss with reference to inventory levels.

decrease −400

6) Why is a level of total output = $6400 billion not an equilibrium? Discuss with reference to inventory levels.

increase +200

Name______________________

Exercise 11.2

1) Fill in the blanks below assuming planned investment = 1700:

Total Income and Output (GDP)	Planned Consumption	Planned Investment	Total Planned Expenditures	Tendency of Output
5600	4000	1700	5700	decrease
6400	4700	1700	6400	no change
7200	5400	1700	6200	increase
8000	6100	1700	7800	increase

2) How much is equilibrium GDP? 6400

3) Now suppose businesses ***increase*** planned investment ***by*** $200 billion. Fill in the table below to reflect the new level of planned investment and its effect on equilibrium:

Total Income and Output (GDP)	Planned Consumption	Planned Investment	Total Planned Expenditures	Tendency of Output
5600	4000	1900	5900	decrease
6400	4700	1900	6600	decrease
7200	5400	1900	7300	increase
8000	6100	1900	8000	no change

4) How much is equilibrium GDP now? 8000

5) Calculate the value of the multiplier (show all calculations):

$$\frac{1600}{200} = 8$$

Name________________________________

Exercise 11.3 ***MATH!***

1) Complete the table below:

Total Income or GDP (Y)	Net Taxes (= taxes - transfers)	Disposable Income (DI)	Consumption (C)
4000	600		3200
5000	600		4000
6000	600		4800
7000	600		5600

2) Calculate the marginal propensity to consume (MPC)

3) Calculate the marginal propensity to save (MPS)

4) Calculate the specific form of the equation for the consumption function from the general equation

$$C = M(DI) + b$$

where C = planned consumption
DI = disposable income
M = slope of the consumption function
b = y intercept of the consumption function

Name________________________________

Exercise 11.4 ***MATH!***

1) Complete the table below assuming:

C=0.8(DI) + 480
I=100
G=600
X=600
IM=300 at all levels of real GDP

GDP	Net Taxes	DI	C	I	G	X	IM	TS	Inventories (+:increase) (-: decrease)
4000	600								
5000	600								
6000	600								
7000	600								

where DI = disposable income
C = consumption
I = investment
G = government purchases
X = exports
IM = imports
TS = total spending

2) At what level of real GDP is the economy at an equilibrium?

3) Why is a level of real GDP = 4000 not an equilibrium? Discuss with reference to inventory levels

4) Why is a level of real GDP = 7000 not an equilibrium? Discuss with reference to inventory levels

Name________________________________

Exercise 11.5 ***MATH!***

Let C = consumption

I = investment

G = government purchases

X = exports

IM = imports

T = net taxes

Y = GDP

All numbers are in billions of dollars

1) C = 100 + 0.8DI

I = 700

G = 500

X = 200

IM = 200

T = 500

Calculate equilibrium GDP

2) C = 200 + 0.8DI

I = 800

G = 500

T = 500

IM = 300

Y = 5000

Calculate the value of exports

Name________________________________

Exercise 12.1

1) The Government wants to conduct ***fiscal policy***. That would involve manipulating which of the following?

Taxes	*yes*	*no*
Interest rates	*yes*	*no*
Money supply	*yes*	*no*
Government spending	*yes*	*no*
Value of the $	*yes*	*no*

2) The government wants to conduct ***expansionary fiscal policy*** to reduce unemployment. This would involve

increasing taxes	*yes*	*no*
decreasing taxes	*yes*	*no*
increasing government spending	*yes*	*no*
decreasing government spending	*yes*	*no*
decreasing interest rates	*yes*	*no*

3) The government wants to conduct ***contractionary fiscal policy*** to fight inflation. This would involve

increasing taxes	*yes*	*no*
decreasing taxes	*yes*	*no*
increasing government spending	*yes*	*no*
decreasing government spending	*yes*	*no*
decreasing interest rates	*yes*	*no*

Name________________________________

Exercise 12.2

Assume I = 840 at all levels of total output
G = 2000 at all levels of total output

Total Output or GDP (Y)	Consumption (C)	Investment (I)	Government Purchases (G)	Total Spending (TS)	Tendency of Output
8000	5760				
9000	6560				
10000	7360				
11000	8160				

1) Complete the table above. (Do not leave any cells blank)

2) At what level of real GDP is the economy at an equilibrium?

Now suppose this level of GDP is so high that it causes inflation.

3) What kind of fiscal policy can the government use to fight the inflation? (Hint: expansionary or contractionary)

4) Should government spending be increased or decreased to fight inflation?

5) Suppose the government decides to ***change*** government spending ***by*** $200 billion in order to ***fight inflation***. Complete the table below using the new level of government spending:

Total Output or GDP (Y)	Consumption (C)	Investment (I)	Government Purchases (G)	Total Spending (TS)	Tendency of Output

6) What is the new equilibrium real GDP?

7) By how much has the equilibrium real GDP changed?

Name______________________________

Exercise 12.3

Assume I = 840 at all levels of total output
G = 1400 at all levels of total output

Total Output or GDP (Y)	Consumption (C)	Investment (I)	Government Purchases (G)	Total Spending (TS)	Tendency of Output
8000	5760				
9000	6560				
10000	7360				
11000	8160				

1) Complete the table above. (Do not leave any cells blank)

2) At what level of real GDP is the economy at an equilibrium?

Now suppose this level of GDP is so low that the economy is in a recession.

3) What kind of fiscal policy can the government use to fight the recession? (Hint: expansionary or contractionary)

4) Should government spending be increased or decreased to fight the recession?

5) Suppose the government decides to ***change*** government spending ***by*** $400 billion in order to ***fight the recession***. Complete the table below using the new level of government spending:

Total Output or GDP (Y)	Consumption (C)	Investment (I)	Government Purchases (G)	Total Spending (TS)	Tendency of Output

6) What is the new equilibrium real GDP?

7) By how much has the equilibrium real GDP changed?

Name________________________________

Exercise 12.4 (Part1) ***CHALLENGE!***

GDP	Net Taxes	DI	C	I	G	X	IM	TS	Inventories (+:increase) (-:decrease)
	100	2000	1600						
	100	2600	2100						
	100	3200	2600						
	100	3800	3100						
	100	4400	3600						

where DI = disposable income
C = consumption
I = investment
G = government purchases
X = exports
IM = imports
TS = total spending

1) Complete the table below assuming:

I=300 at all levels of real GDP	X=100 at all levels of GDP
G=400 at all levels of GDP	IM=200 at all levels of GDP

2) At what level of GDP is the economy at an equilibrium?

3) What is the size of the budget deficit?

Name________________________________

Exercise 12.4 (Part2) ***CHALLENGE!***

1) Suppose the government decides to increase the level of equilibrium GDP by *increasing* government purchases (G) by $200 billion. Complete the table below with the new level of (G)

GDP	Net Taxes	DI	C	I	*New G*	X	IM	TS	Inventories (+:increase) (-:decrease)
	100	2000	1600						
	100	2600	2100						
	100	3200	2600						
	100	3800	3100						
	100	4400	3600						

2) At what level of GDP is the economy at an equilibrium now?

3) How much did equilibrium GDP increase when G was increased by $200 billion?

4) Calculate the value of the multiplier showing all your calculations:

a) Using change in equilibrium GDP formula

b) Using the MPC formula

5) What is the size of the budget deficit now?

6) When the government conducts expansionary fiscal poilicy, does the budget deficit increase or decrease?

Name____________________________________

Exercise 12.5 ***INTERNET!***

Visit the National Budget Simulation at:

http://garnet.berkeley.edu:3333/budget/budget.html

You have been appointed as top Economic Advisor to the President. Your task is to balance the budget in ONE YEAR! Set about your task seriously. Once you attain your goal or get the closest to your goal that you possibly can (make the deficit or surplus as small as possible), document your changes below:

1) What is the size of the deficit or surplus you are left with?

2) List the categories that you created changes in and the amounts:

SPENDING

TAX EXPENDITURES

Name________________________________

Exercise 12.6 ***MATH!***

Assume a MPC of 3/4 or 0.75. The government wants to conduct expansionary fiscal policy. Specifically, the government wants to increase equilibrium GDP by $800 billion. Answer the following based on this idea.

1) How much of an increase in *government spending* (G) will be required to achieve the $800 billion increase in GDP?

2) If instead of increasing government spending, *investment* (I) were to increase, how much would investment have to increase to create the $800 billion increase in GDP?

3) If investment and government spending did not change, how much would *exports* (X) have to increase to create the $800 billion increase in GDP?

4) Now consider this carefully. Suppose none of the above changed. If the government were to *cut taxes* in order to increase GDP by $800 billion, how much of tax cuts would be required?

Name________________________________

Exercise 12.7 ***MATH!***

MPC=0.795
Potential GDP = $6000 billion
Equilibrium GDP = $6500 billion

1) Is there a recessionary gap or an inflationary gap in the situation above?

2) What is the ***size*** of the gap?

3) By how much would government spending have to change to get rid of the gap?

4) Would government spending be increased or decreased?

5) If government spending is changed, is that fiscal policy or monetary policy?

6) In this case is the policy expansionary or contractionary?

7) If taxes are changed instead of government spending, by how much would taxes have to be changed?

8) If taxes are changed, is that fiscal policy or monetary policy?

9) In this case is the policy expansionary or contractionary?

10) Are taxes increased or decreased?

11) In dollar amounts, does it take *less* in terms of government spending or taxes to close the gap?

Name________________________________

Exercise 12.8 ***MATH!***

Let

C = consumption

I = investment

G = government purchases

X = exports

IM = imports

T = net taxes

Y = GDP

All numbers are in billions of dollars

$C = 100 + 0.8DI$

$I = 900$

$G = 500$

$X = 200$

$IM = 400$

$T = 500$

1) Calculate equilibrium GDP (Y)

2) If G is increased by $200 billion, calculate the new equilibrium GDP (Y)

3) If instead of a change in G, T is decreased by $200 billion, calculate the new equilibrium GDP (Y)

4) Calculate the value of the multiplier for changes in G and the multiplier for changes in T

Name________________________________

Exercise 12.9 ***MATH!***

1) If MPC = 0.74, and government spending ***increased*** by $2,500 billion, by how much would GDP increase?

2) Suppose the MPC is the same as before but there was a **balanced budget amendment** in place requiring a balanced budget at all times. ***Would taxes have to be increased or decreased*** along with the changes in government spending in question 1?

3) By how much would the balanced budget law require taxes to be changed?

4) How much will GDP change as a result of the change in taxes?

5) What is the ***combined*** effect on GDP of the change in ***both*** government spending ***and*** taxes?

Name________________________________

Exercise 13.1

John had been saving pennies and dimes for years. One day when he counted them all, he realized he had saved $900.80. The next day John deposited this money into his checking account at Bank of USA. Assume reserve requirement = 20%.

1) In the balance sheet below show the changes on the Bank USA balance sheet when John initially deposits his money:

Assets	Liabilities

2) In the balance sheet below, show the changes on the Bank USA balance sheet after the bank is "loaned out," that is, after it has made loans to its maximum potential.

Assets	Liabilities

3) How many dollars worth of new loans can Bank USA make?

4) How many dollars worth of new loans can the banking system as a whole make?

Name________________________________

Exercise 13.2

1) Draw a money supply and money demand graph below with the *quantity of money* on the x-axis and the *interest rate* on the y-axis.

(A)

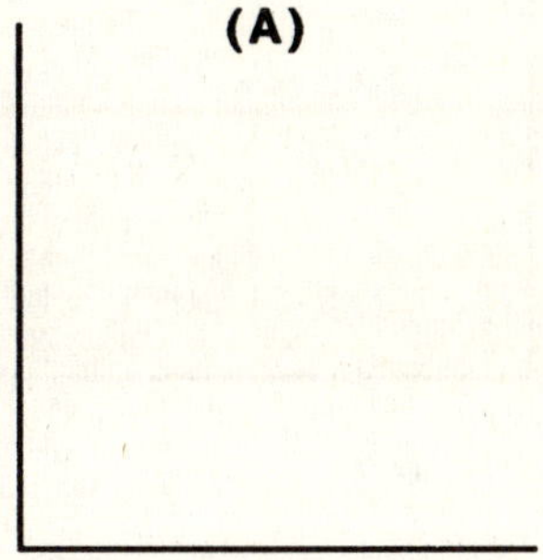

2) If the Federal Reserve (or FED) ***sells*** government securities, show what happens to money supply on graph (A) above.

3) Is that an increase or decrease in money supply?

4) Does the interest rate increase or decrease?

5) What happens to spending and GDP?

6) Is this fiscal policy or monetary policy?

7) Is this an expansionary policy or a contractionary policy?

8) Draw a money supply and money demand graph below with the *quantity of money* on the x-axis and the *interest rate* on the y-axis.

(B)

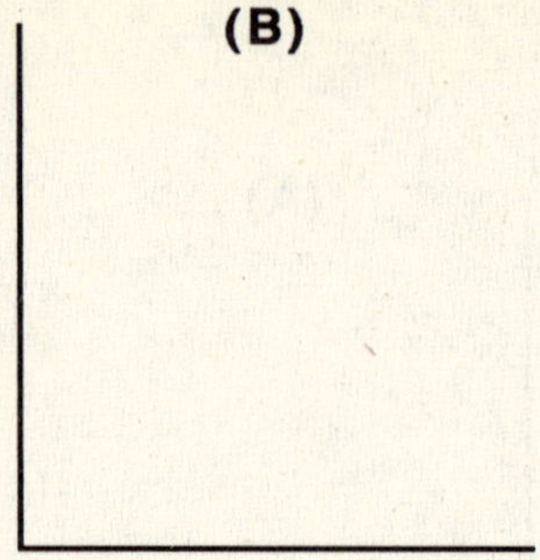

9) If the Federal Reserve (or FED) ***buys*** government securities, show what happens to money supply on graph (B) above.

10) Is that an increase or decrease in money supply?

11) Does the interest rate increase or decrease?

12) What happens to spending and GDP?

13) Is this fiscal policy or monetary policy?

14) Is this an expansionary policy or a contractionary policy?

Name________________________________

Exercise 13.3 ***MATH!***

For the following questions, assume that all banks in the system are loaned up (that is, they have no excess reserves):

1) If the reserve ratio or reserve requirement is 7.5%, and the FED ***buys*** $2,700 worth of government securities from the public

a) Does money supply increase or decrease?

b) By how much does money supply change?

2) If the FED ***sells*** government securities worth $2,700 to the public

a) Does money supply increase or decrease?

b) By how much does money supply change?

3) If the reserve ratio is 9.25% and the FED wants to ***increase*** money supply by $20,000

a) Does it buy or sell government securities to the public?

b) How many dollars worth of government securities does it buy or sell?

4) If the reserve ratio is 9.25% and the FED wants to ***decrease*** money supply by $20,000

a) Does it buy or sell government securities to the public?

b) How many dollars worth of government securities does it buy or sell?

Name________________________________

Exercise 14.1

In each of the graphs below, show how a change in the particular factor or event presented will affect the **aggregate demand curve**. Label the new aggregate demand curve clearly and show the direction of the change.

a) The government increases its spending on goods and services.

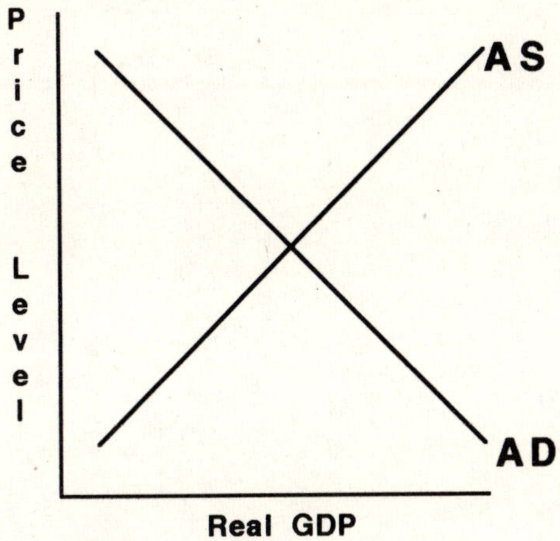

What happens to equilibrium GDP?

What happens to the price level?

b) Households and businesses become more optimistic about the future.

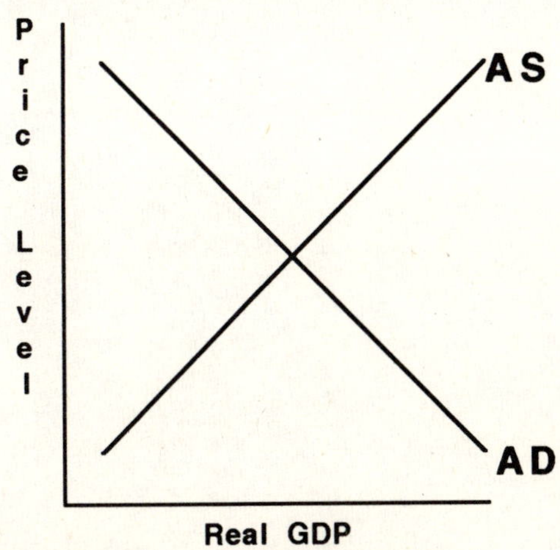

What happens to equilibrium GDP?

What happens to the price level?

c) There is a sudden drop in the stock market and the average price of stock declines.

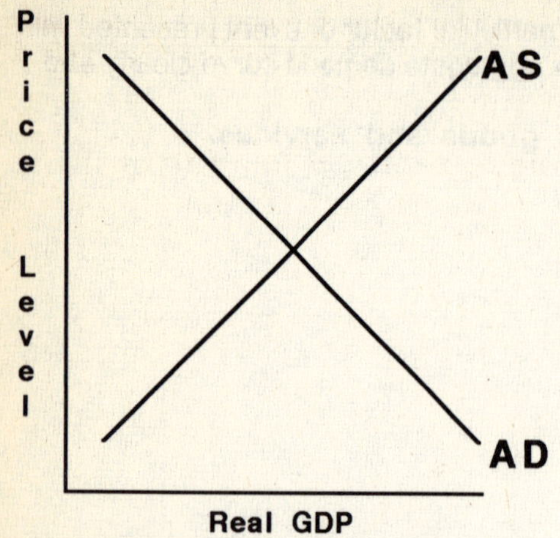

What happens to equilibrium GDP?

What happens to the price level?

d) Average incomes in foreign countries increase along with foreign price levels.

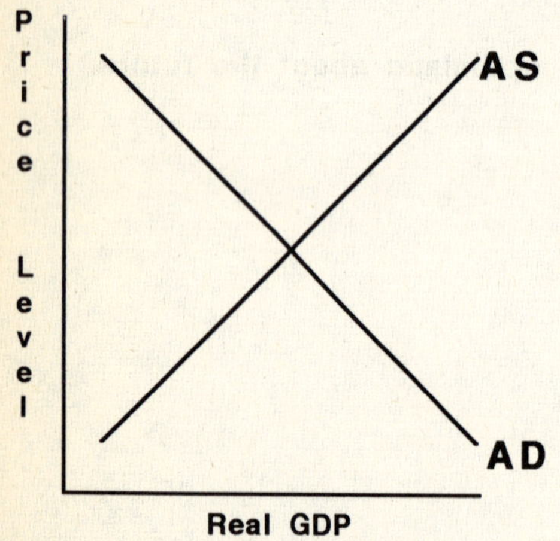

What happens to equilibrium GDP?

What happens to the price level?

Name____
Exercise 14.2

In each of the graphs below, show how a change in the particular factor or event presented will affect the **aggregate supply curve**. Label the new aggregate supply curve clearly and show the direction of the change.

a) There is an increase in the stock of capital equipment and the size of the labor force.

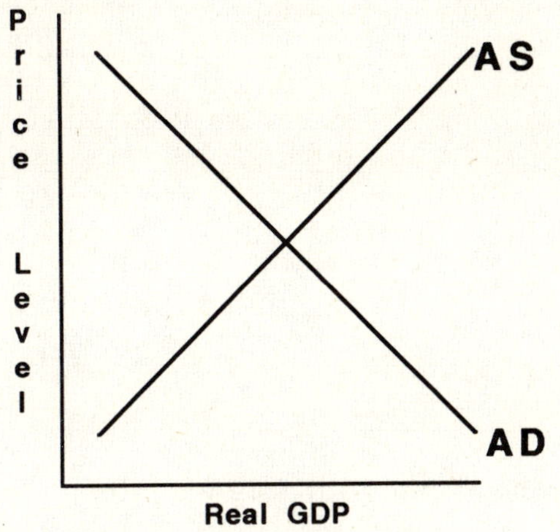

What happens to equilibrium GDP?

What happens to the price level?

b) The average wage rate in the economy increases.

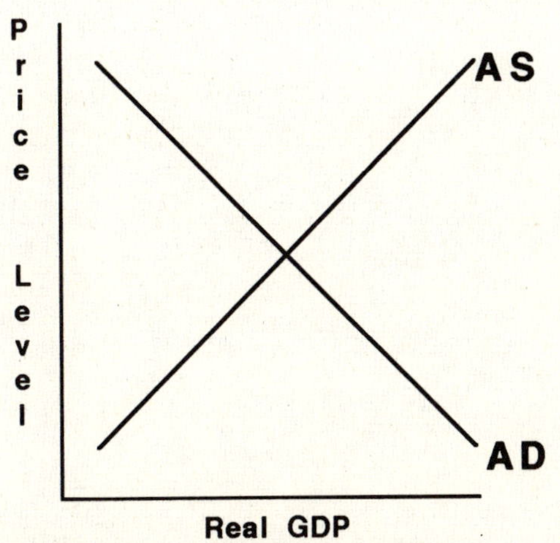

What happens to equilibrium GDP?

What happens to the price level?

c) The labor force becomes more educated resulting in higher labor productivity.

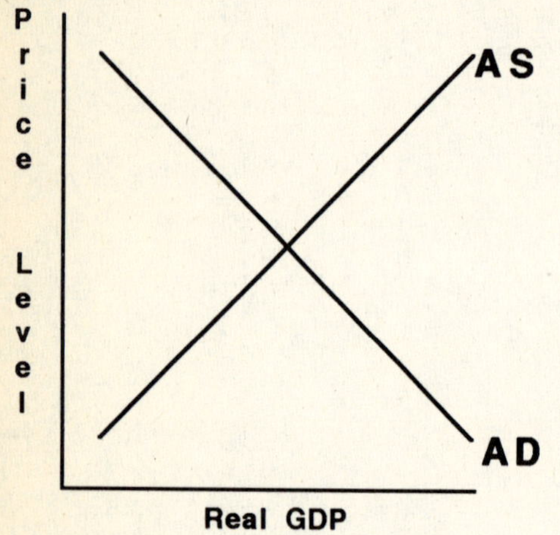

What happens to equilibrium GDP?

What happens to the price level?

d) The price of key raw materials such as steel and lumber increases.

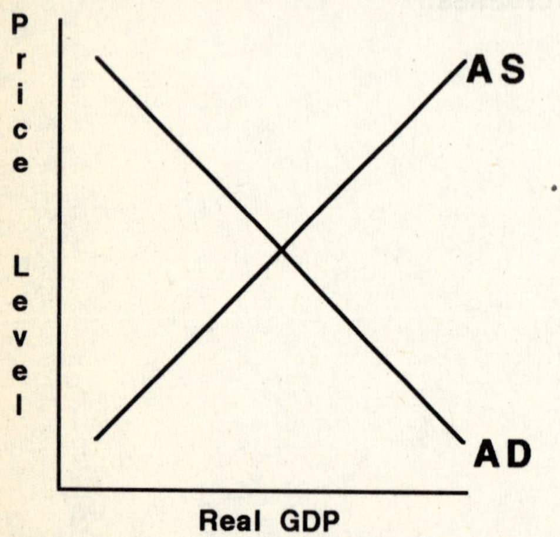

What happens to equilibrium GDP?

What happens to the price level?

Name________________________________

Exercise 15.1

1) In the graphs below use a new AS curve to show the prediction of the theory of rational expectations in the face of an anticipated change in economic policy:

a) Expansionary economic policy is undertaken

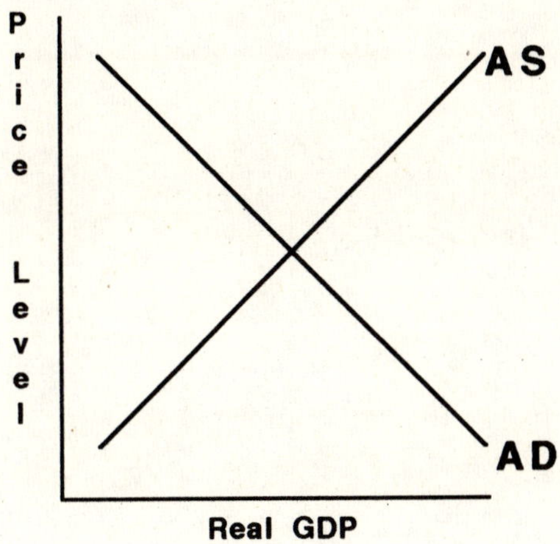

b) Contractionary economic policy is undertaken

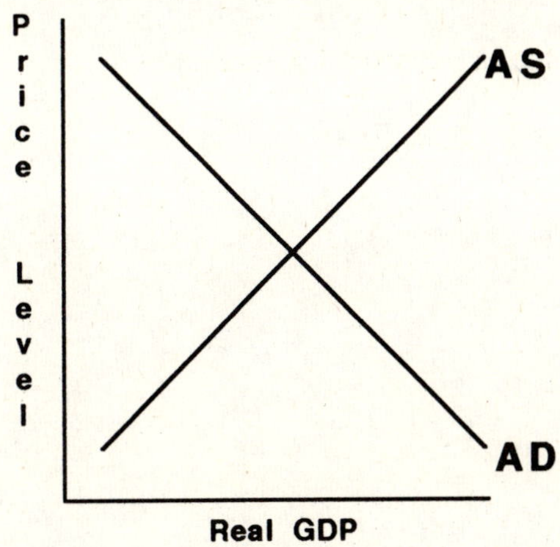

2) In the graph below show the prediction of the supply-side economics school in the face of the following economic policies:

a) A decrease in the marginal tax rates.

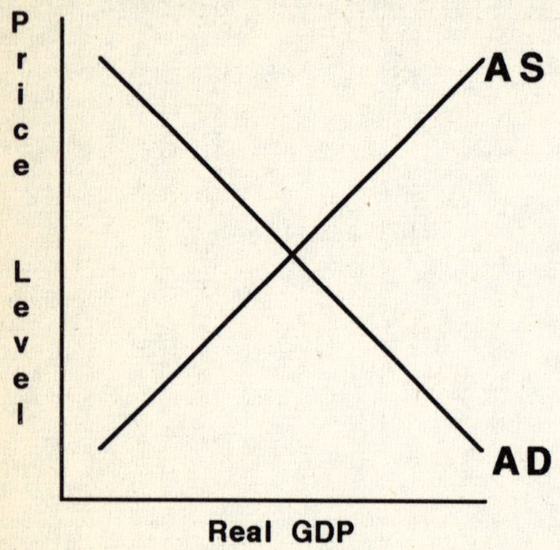

b) A reduction in government regulations.

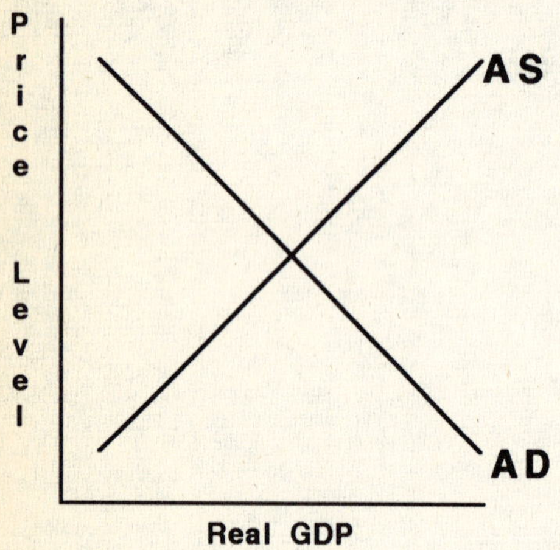

Name________________________________

Exercise 15.2 ***INTERNET!***

Visit the Bank of America at:

http://www.bankamerica.com/econ_indicator/econ_indicator.html

Write down a summary of the current economic conditions including a brief explanation of the economic statistics released this week.

Name______________________________

Exercise 16.1

Production possibilities per unit of economic resources:

	Cars	VCRs
United States	900	300
Brazil	200	180

1) Which country has an absolute advantage in the production of cars?

2) Which country has an absolute advantage in the production of VCRs?

3) Fill in the following table:

	Opportunity Cost of a Car	Opportunity Cost of a VCR
United States		
Brazil		

4) Which country has a comparative advantage in the production of cars?

5) Which country has a comparative advantage in the production of VCRs?

6) Which of the two products should the United States specialize in producing?

7) Which of the two products should Brazil specialize in producing?

Name________________________

Exercise 16.2 (Part1)

The following table contains production possibilities assuming that Thailand and Taiwan both have 100 units of economic resources equally suited to the production of TVs and shoes:

	TVs	Shoes (pairs)
Taiwan	100	400
Thailand	250	500

Note: *Taiwan can produce either 100 TVs or 400 pairs of shoes.*
Thailand can produce either 250 TVs or 500 pairs of shoes.

1) Graph the production possibilities curve for Taiwan and Thailand:

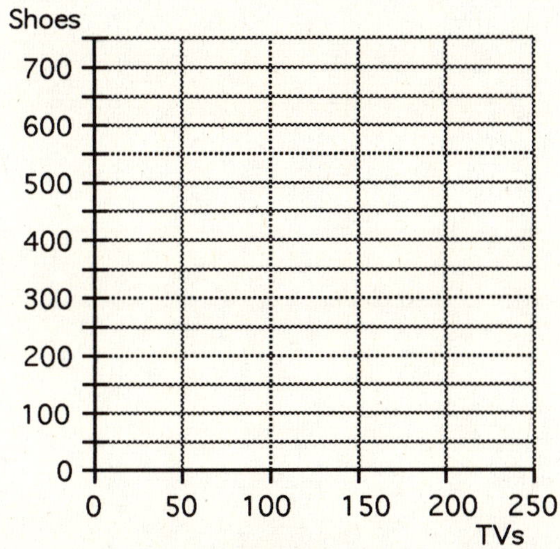

TAIWAN

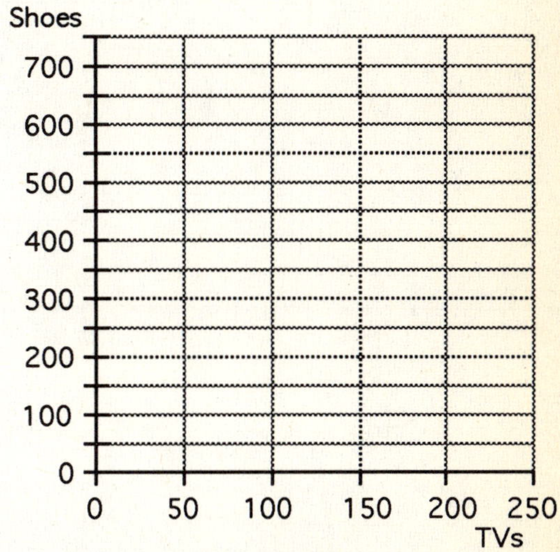

THAILAND

2) If Taiwan produces 100 TVs, what is the maximum number of pairs of shoes it can produce? (Mark this point on the graph for Taiwan with an **A**.)

3) If Thailand produces 300 pairs of shoes, what is the maximum number of TVs it can produce? (Mark this point on the graph for Thailand with a **B**.)

Name______________________________

Exercise 16.2 (Part2)

1) Calculate the following and fill the table:

	Opportunity cost of a TV	Opportunity cost of a pair of shoes
Taiwan		
Thailand		

Suppose the two countries now specialize in the area of their comparative advantage.

Taiwan will specialize in:

Thailand will specialize in:

Assume the following exchange ratio is agreed on:

3 pairs of shoes = 1 TV

2) If Taiwan exports 300 pairs of shoes to Thailand, how many TVs will it get in return?

3) Citizens of Taiwan will be able to consume the following quantities of products *after trading*:

Pairs of Shoes	TVs

Mark this consumption point with a **C** on the graph for Taiwan in Part 1 of this exercise.

4) Was this level of consumption possible *before* trade?

5) Citizens of Thailand will be able to consume the following quantities of products *after trading*:

Pairs of Shoes	TVs

Mark this consumption point with a **D** on the graph for Thailand in Part 1 of this exercise.

6) Was this level of consumption possible *before* trade?

[illegible]

1) [illegible]

[illegible]	Opportunity cost of [illegible]	Opportunity cost of [illegible]
[illegible]		
[illegible]		

Suppose the two countries now specialize in the area of their comparative advantage.

[illegible]

[illegible]

Answer the following questions [illegible]

[illegible]

2) [illegible]

3) [illegible]

[illegible]	[illegible]

[illegible]

[illegible]

4) [illegible]

[illegible]	[illegible]

[illegible]

[illegible]

Name________________________________

Exercise 17.1

In each of the graphs below, show a **shift** in **either** the **demand curve** ***or*** the **supply curve** or **both** as a result of the event presented. Label the new supply and / or demand curve clearly and show the direction of the change.

a) Japan experiences 2% inflation while inflation in the US is 7%.

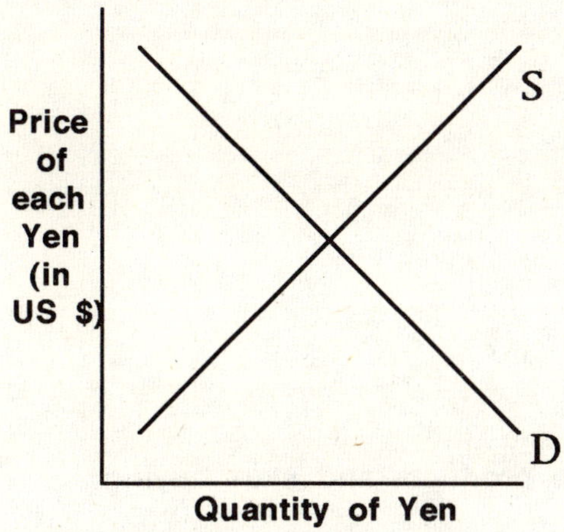

1) What happens to the Japanese yen?

2) What happens to the US dollar?

b) Average income in the US decreases as the US economy experiences a recession.

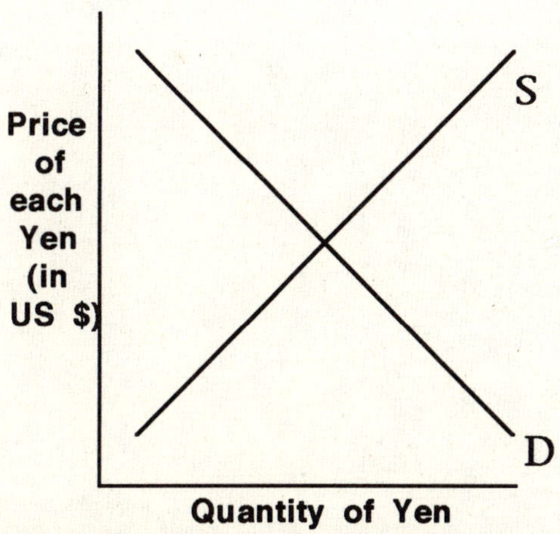

3)What happens to the Japanese yen?

4) What happens to the US dollar?

c) Interest rates in Japan are made lower than in the US.

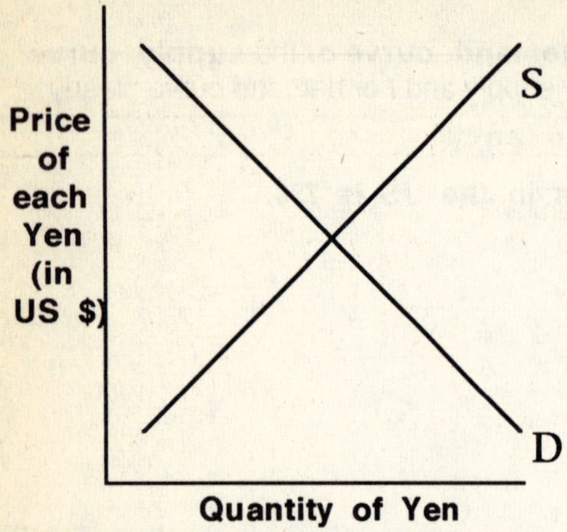

5) What happens to the Japanese yen?

6) What happens to the US dollar?

d) Japanese consumers start prefering US made products and start buying them.

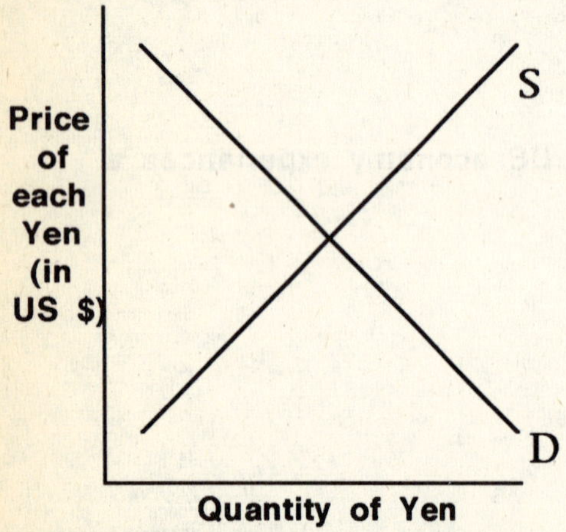

7) What happens to the Japanese yen?

8) What happens to the US dollar?

Name___________________________________

Exercise 17.2 ***INTERNET!***

Visit the Universal Currency Converter at:

http://www.xe.net/currency/

1) How many Indian rupees would exchange for one US dollar?

2) How many German marks would exchange for one US dollar?

3) How many British pounds would exchange for one US dollar?

4) 1000 Japanese yen would exchange for how many Danish kroner?

SECTION TWO

Mathematical Economics

Topic 1 Percentages

A percent figure is a figure expressed per *century* or per hundred. For example, if you scored 22 points on a quiz out of a maximum possible 30 points, your percent score is

$$\frac{22}{30} \times 100$$

$$= \mathbf{73.33\%}$$

Getting 22 points out of maximum 30 is the same as getting 73.33 points out of a maximum 100.

If a quiz is out of 45 points and you want to earn a 90% score on that quiz, how many points do you need to get?

We can calculate the answer easily by remembering that the word **of** in mathematics always means **multiply by**.

So 90% of 45 points

$= 0.9 \times 45$

$= 40.5$ points.

You need to get 40.5 points on the quiz to get a 90% score.

Test yourself: If you borrow $900 at an interest rate of 15% per year, how much would you have to pay in interest during the first year?

Topic 2 Percentage Change

The percentage change formula is:

$$\% \text{ change} = \frac{\text{new value} - \text{old value}}{\text{old value}} \times 100$$

For example, if the Consumer Price Index is as follows:

Years	Consumer Price Index
Year 1	144.5
Year 2	148.2

The % change in the Consumer Price Index between Year 1 and Year 2 is calculated as follows:

$$\% \text{ change} = \frac{148.2 - 144.5}{144.5} \times 100$$

= ___________%
(please calculate)

Topic 3 Equation of a Straight Line

The equation of a straight line describes the mathematical relationship between two variables. The general form of the equation of a straight line is:

$$Y = M(X) + b$$

where X and Y are the two variables
M is the slope
b is the y-intercept

When the line is drawn as a graph we observe the following:

a) variable Y is plotted on the vertical axis
b) variable X is plotted on the horizontal axis
c) M is the slope as calculated by the formula below
d) b is where the line intersects the Y axis

The slope is defined as follows:

$$\text{slope} = \frac{\text{change in Y}}{\text{change in X}}$$

If the calculated value of the slope is positive, it means that X and Y have a direct relationship. If the slope is equal to +5 it means that when X **increases** by one unit, Y **increases** by 5 units.

If the calculated value of the slope is negative, it means that X and Y have an inverse relationship. If the slope is equal to -5 it means that when X **increases** by one unit, Y **decreases** by 5 units.

Topic 4 Supply and Demand

A. Introduction

The demand curve can be represented by the function:

$$Qd = f(P)$$

where Qd is quantity demanded and P is the price. This equation is read as "Qd equals f of P." This means that the quantity demanded depends on the price of the product.

The supply curve can be represented by the function:

$$Qs = f(P)$$

where Qs is the quantity supplied and P is the price.

If the demand curve is a straight line, the function above takes the form

$$Qd = M_d(P) + b_d$$

or

$$Qd = b_d + M_d(P)$$

where M_d is the (negative) slope of the demand curve and b_d is the y-intercept.

The straight line supply curve takes the form

$$Qs = M_s(P) + b_s$$

or

$$Qs = b_s + M_s(P)$$

where M_s is the (positive) slope of the supply curve and b_s is the y-intercept.

Consider these specific forms for the demand and supply equations:

$$Qd = 4000 - 250P$$
$$Qs = 1000 + 500P$$

Our first task will be to find equilibrium price and quantity.

B. Equilibrium

STEP 1: Find equilibrium price

Equilibrium is reached where the demand and supply curves cross each other. So our equilibrium condition is

$$Qd = Qs$$

Therefore

$$4000 - 250P = 1000 + 500P$$
$$3000 = 750P$$
$$P = 3000/750$$
$$P = 4$$

The equilibrium price is \$4.

STEP 2: Find equilibrium quantity

Replace the value of equilibrium price (\$4) in *either* the demand or the supply curve. Both give us the same answer.

$$Qd = 4000 - 250(4)$$
$$Qd = 3000$$

$$Qs = 1000 + 500(4)$$
$$Qs = 3000$$

Equilibrium quantity is 3000.

C. Increase and Decrease in Demand

Our original demand curve is

Qd = 4000 - 250P

An increase in consumer income leads to an increase in demand in the case of normal goods. Let us assume our demand equation represents demand for a normal good. If income increases, demand will increase. The new equation may look like this:

Qd = 4000 - 250P + **750**

or Qd= 4750 - 250P

What effect will this increase in demand have on equilibrium price and quantity? We need to use the two-step process described above to find the new equilibria.

STEP 1: Find equilibrium price

Set Qd = Qs

or 4750 - 250P = 1000 + 500P

3750 = 750P

P = 5

STEP 2: Find equilibrium quantity

Replace the value of P found in step 1 in either demand or supply equations.

Qd = 4750 - 250(5)

Qd = 3500

Go ahead and replace the value of P in the supply equation. You should get the same answer. Equilibrium quantity is 3500

An increase in income has caused equilibrium price to increase from $4 to $5 and equilibrium quantity to increase from 3000 to 3500.

Test your understanding: Calculate the new equilibrium price and quantity if income *decreased* and the new demand equation was

Qd = 4000 - 250P - **750**

C) Price Ceilings

The supply and demand equations we started off with are

Qd = 4000 - 250P
Qs = 1000 + 500P

If a price ceiling is to be effective it must be placed below equilibrium. We found equilibrium price to be $4. Let us suppose that a price ceiling is placed at P=$2.

Then,

Qd = 4000 - 250(2)
Qd = 4000 - 500
Qd = 3500

And,

Qs = 1000 + 500(2)
Qs = 1000 + 1000
Qs = 2000

Since Qd > Qs, a shortage exists. The amount of the shortage is
3500 - 2000 = 1500

D. Price Floors

For a price floor to be effective, it must be placed above equilibrium. Recall that our equilibrium price is $4. Let us suppose that a price floor is placed at P = $6.

Then,

Qd = 4000 - 250(6)
Qd = 4000 - 1500
Qd = 2500

And,

Qs = 1000 + 500(6)
Qs = 1000 + 3000
Qs = 4000

Since Qs > Qd, a surplus exists. The amount of the surplus is
4000 - 2500 = 1500

Topic 5 Consumption

Consider a table showing various levels of income (or GDP), net taxes, and the associated levels of consumption:

Total Income or GDP (Y)	Net Taxes (= taxes - transfers)	Disposable Income (DI)	Consumption (C)
4000	200		3600
4500	200		4050
5000	200		4500
5500	200		4950

Disposable income is the total income after subtracting taxes and adding in government transfer payments. In short,

DI = Total Income - taxes + transfers

Since Total Income = GDP and net taxes = (taxes-transfers),

DI = GDP - net taxes

Fill in the table above using this formula.

From a table like this, it is possible to derive the equation of the line for the consumption function.

The general form of any consumption function is

C = M(DI) + b
where M = slope
and b = y intercept

To find the specific form that the consumption function takes in this table of figures above, follow these 3 steps:

STEP 1: Find Slope (M)

M is the slope of the consumption function, which we call the marginal propensity to consume or MPC.

$$M = MPC = \Delta C / \Delta DI$$

We derive the change in C by calculating the difference between two consecutive levels of C (example, 4050 - 3600 = 450).

We derive the change in DI by calculating the difference between two consecutive levels of DI (example, 4300 - 3800 = 500).

So M = MPC = 450/500 = **0.9**

STEP 2: Find Y intercept (b)

Since M = 0.9,

$$C = 0.9(DI) + b$$

Replacing values for C and DI from any row in the table above,

$$3600 = 0.9(3800) + b$$
$$3600 = 3420 + b$$
$$b = 3600 - 3420$$
$$b = \mathbf{180}$$

STEP 3: Replace values of M and b in the general form of the equation to get the specific form of the equation

$$\mathbf{C = 0.9(DI) + 180}$$

This equation for the consumption function will give you the value for C for any value of DI. Try plugging in one of the values for DI in the table. You should get the same value for C as the table. Now this is useful because it works not only for all the DI values in the table but for *any* DI that you wish to plug in. Given any DI, we can calculate the corresponding level of C.

Topic 6 Macroeconomic Equilibrium

Our goal here is to find mathematically, the level of equilibrium GDP.

Let us now use the specific form of the consumption function we derived in the previous topic:

$$C = 0.9(DI) + 180$$

To keep matters simple, let us assume that net taxes (T), investment (I), government purchases (G), exports (X) and imports (IM) and are constants or fixed. Assume that the fixed values for these variables are:

$$I = 900$$
$$G = 500$$
$$T = 500$$
$$X = 200$$
$$IM = 300$$

To find equilibrium GDP (Y), start with the consumption function:

$$C = 0.9(DI) + 180$$

Since $DI = Y - T$,

$$C = 0.9(Y-T) + 180$$

Since T = 500,

$$C = 0.9(Y-500) + 180$$
$$C = 0.9Y - 450 + 180$$
$$C = 0.9Y - 270$$

Now,

$$TS = C + I + G + (X - IM)$$

where TS = total spending

Replacing C in this equation with 0.9 - 270,

$$TS = 0.9Y - 270 + 900 + 500 + (200 - 300)$$
$$TS = 0.9Y + 1030$$

The equilibrium condition is GDP (Y) = Total Spending (TS)

So,

$$Y = 0.9Y + 1030$$
$$Y - 0.9Y = 1030 \text{ (by subtracting 0.9Y from both sides)}$$
$$0.1Y = 1030$$
$$Y = 1030/0.1 \text{ (by dividing both sides by 0.1)}$$
$$Y = 10300$$

The equilibrium GDP (Y) is 10,300 for the given consumption function and given values for the other variables: net taxes (T), investment (I), government purchases (G), exports (X) and imports (IM).

OPTIONAL CHALLENGE TOPIC

We can also easily derive the value of equilibrium GDP for any set of numbers in our equations.

The general form of the consumption function is:

$$C = M(DI) + b$$
$$C = M(Y-T) + b$$

The general form of total spending is:

$$TS = C + I + G + (X - IM)$$

So,

$$TS = MY - MT + b + I + G + (X - IM)$$

The equilibrium condition is:

$$Y = TS$$
$$Y = MY - MT + b + I + G + (X - IM)$$
$$Y - MY = -MT + b + I + G + (X - IM)$$
$$Y(1-M) = -MT + b + I + G + (X - IM)$$

Therefore,

$$Y = \frac{-MT + b + I + G + (X - IM)}{(1-M)}$$

See if you can use this formula to get the equilibrium GDP in the example above.

Let us start off with same levels of total spending (TS) in the economy as did before:

$C = 0.9(DI) + 180$
$I = 900$
$G = 500$
$T = 500$
$X = 200$
$IM = 300$

Now suppose the government wants to conduct expansionary fiscal policy and increases G by 200. That would increase G from the original 500 up to 700. Now total spending (TS) in the economy would be:

$C = 0.9(DI) + 180$
$I = 900$
$G = 700$
$T = 500$
$X = -100$
$IM = 300$

This change in G will now create a ***new*** equilibrium GDP.

Start with the consumption function:

$C = 0.9(DI) + 180$

$C = 0.9(Y-T) + 180$

$C = 0.9(Y-500) + 180$

$C = 0.9Y - 450 + 180$

$C = 0.9Y - 270$

Now,

$TS = C + I + G + (X - IM)$

Replacing C in this equation with 0.9 - 270,

$TS = 0.9Y - 270 + 900 + \mathbf{700} + (200 - 300)$
$TS = 0.9Y + 1230$

The equilibrium condition is:

GDP (Y) = Total Spending (TS)

The equilibrium condition is:

GDP (Y) = Total Spending (TS)

So,

$Y = 0.9Y + 1230$
$Y - 0.9Y = 1230$ (by subtracting 0.9Y from both sides)
$0.1Y = 1230$
$Y = 1230/0.1$ (by dividing both sides by 0.1)
$Y = 12300$

The new equilibrium GDP (Y) is 12,300 for the given consumption function and given values for the other variables including the new value for G.

Now compare the previous equilibrium GDP (10,300 from the previous topic) to the new equilibrium GDP (12,300). The increase in G by 200 caused an increase in equilibrium GDP by 2000! This is known as the multiplier effect. We can calculate the multiplier as follows:

$$\text{multiplier} = \frac{\text{change in equilibrium GDP } (\Delta Y)}{\text{change in total spending } (\Delta TS)} = \frac{2000}{200} = 10$$

A change in *any* of the components of total spending (C, I, G, X, or IM) will result in ten times as much of a change in equilibrium GDP! (How will the effect of a change in IM be different from the others?)

The multiplier can also be calculated using the concept of the marginal propensity to consume (MPC):

$$\text{multiplier} = \frac{1}{1-MPC}$$

Since the consumption function is:

$C = 0.9(DI) + 180$

the $MPC = 0.9$.

therefore,

$\text{multiplier} = 1/(1 - 0.9) = 10$

Either formula gives us the same answer for the multiplier.

We can summarize the relationship between changes in the equilibrium GDP and total spending (TS) with the following equation:

Δ GDP = ΔTS x multiplier

If the government wants to raise GDP by $500 billion and the multiplier is 10,

Δ GDP = ΔTS x multiplier
500 = ΔTS x 10
ΔTS = 500/10
ΔTS = 50

The government must increase TS must by $50 billion to increase GDP by $500 billion. The $50 billion increase in TS can be achieved not only by an increase in **G**, but also by either:

a) an increase in **C** by $50 billion
b) an increase in **I** by $50 billion
c) an increase in **X** by $50 billion
d) a *decrease* in **IM** by $50 billion

For example, if MPC = 0.75 and net exports (X) increased by $30 million, we can calculate the resulting effect on equilibrium GDP this way:

Firstly, Multiplier = 1/(1-MPC)
= 1/(1-.75)
= 1/.25
= 4

Secondly, using the multiplier formula,

Δ GDP = ΔTS x multiplier
= 30 x 4
= 120

So equilibrium GDP increases by $120 million when exports increase by $30 million and MPC = 0.75.

When taxes are changed, the situation is slightly more complicated. In the case of C, I, G, X, and IM, a $100 change in any one of them leads to a $100 change in total spending (TS). But changing net taxes (T) by $100 does not change TS by $100! Remember this: If MPC = 0.75, then a $100 decrease in taxes will raise disposable income by $100 but that will lead to only a $75 increase in C.

So,

$$\Delta TS = -\Delta T \times MPC$$

If MPC = 0.75, and the government wants to increase GDP by $500 billion, how much must it raise TS?

$$\begin{aligned} \Delta GDP &= \Delta TS \times \text{multiplier} \\ 500 &= \Delta TS \times 4 \\ \Delta TS &= 500/4 \\ &= 125 \end{aligned}$$

How much must taxes change to change TS by 125?

$$\begin{aligned} \Delta TS &= -\Delta T \times MPC \\ 125 &= -\Delta T \times 0.75 \\ \Delta T &= -\ 125/0.75 \\ &= -\ \$166.67 \end{aligned}$$

Taxes have to be decreased by $166.67 in order to increase GDP by $500 billion.

OPTIONAL CHALLENGE TOPIC

We can mathematically derive the formula of the multiplier. From the previous topic we know that the formula for equilibrium GDP is:

$$Y_1 = \frac{-MT + b + I + G + (X - IM)}{(1 - M)}$$

If I were to change by ΔI, the new GDP would be:

$$Y_2 = \frac{-MT + b + I + G + (X - IM) + \Delta I}{(1 - M)}$$

$$\Delta GDP = Y_2 - Y_1$$

$$\Delta Y = \frac{-MT + b + I + G + (X - IM) + \Delta I}{(1 - M)} - \frac{\{-MT + b + I + G + (X - IM)\}}{(1 - M)}$$

$$\Delta Y = \frac{\Delta I}{(1 - M)}$$

So,

$$\frac{\Delta GDP}{\Delta I} = \frac{1}{(1 - M)}$$

Recall that,

$$\text{multiplier} = \frac{\text{change in equilibrium GDP } (\Delta Y)}{\text{change in total spending } (\Delta TS)}$$

Since M = MPC, this is the multiplier formula. Make sure you understand that this formula can be derived not only for ΔI, but also for ΔG, ΔX and ΔIM, remembering that when a ΔIM is considered, the formula will end up with a negative sign before it.

Topic 8 Money and the FED

The formula below describes the maximum potential amount of new money supply created from the open market operations conducted by the Federal Reserve (FED).

Δ Money supply = money value of government securities bought or sold x money multiplier.

money multiplier = 1/required reserve ratio

If required reserve ratio is 10%, money multiplier = 1/.10 = 10

If the Fed buys $55 billion worth of government securities, money supply will increase by (a maximum of) 55 x 10 = $550 billion.

What would happen to the money supply if required reserve ratio was 12% and the FED sold $25 billion worth of government securities?

SECTION THREE

Economic Statistics

Gross Domestic Product (GDP)

INTERNET!
http//www.bea.doc.gov/

The GDP is the economic report card of the country. It measures the total money value of all final products and services produced within the country. The Bureau of Economic Analysis of the US Department of Commerce publishes GDP figures for every quarter as well as for the entire year.

It is important to keep certain things in mind. GDP measures the *money values* of products and services produced. If 100,000 haircuts are given during the quarter or year and each haircut costs $15 on average, then $1,500,000 will be added to GDP. If 1,000,000 lb.. of apples are produced and sold at a price of $2 on average, then $2,000,000 will be added to GDP.

GDP only considers the money values of *final* products and services. Economists distinguish between *final* and *intermediate* products and services. An intermediate product is produced by one firm and sold to another for further processing or to be used as a component in something else. Tires sold by Goodyear to the Ford Motor Co. to be installed in the new cars produced are an example of intermediate products. If Ford paid Goodyear $200 for the tires and then sold the car for $20,000, the cost of the tires is already accounted for in the price of the car. So we add the $20,000 to GDP but do not *also* add $200 *more* for the tires produced. Doing that would result in *double counting.*

The GDP of the United States consists of everything produced *within* the country, regardless of who owns the resources used in production. Production of cars at Honda's plant in Maryville, Ohio is part of our GDP. Production of GM cars in Mexico is part of Mexico's GDP but not our GDP.

Economists often convert GDP figures (also called nominal GDP) into *real GDP.* Real GDP removes the effect of inflation from GDP figures. Two consecutive quarters of decline in the real GDP means that the economy is in a *recession.*

There are two ways of calculating GDP. The *expenditure* approach and the *income* approach. They give the same results because every payment (expenditure) is also a receipt (income). When you spend $100 on a boom box (expenditure), the store gets your $100 (income).

Quarterly GDP figures are published in the Wall Street Journal, Economic Indicators, and the Survey of Current Business. The GDP HOTLINE is: (202) 606-5306. The Internet address for the Bureau of Economic Analysis is at the top of this section.

The following are GDP figures compiled by the Bureau of Economic Analysis.

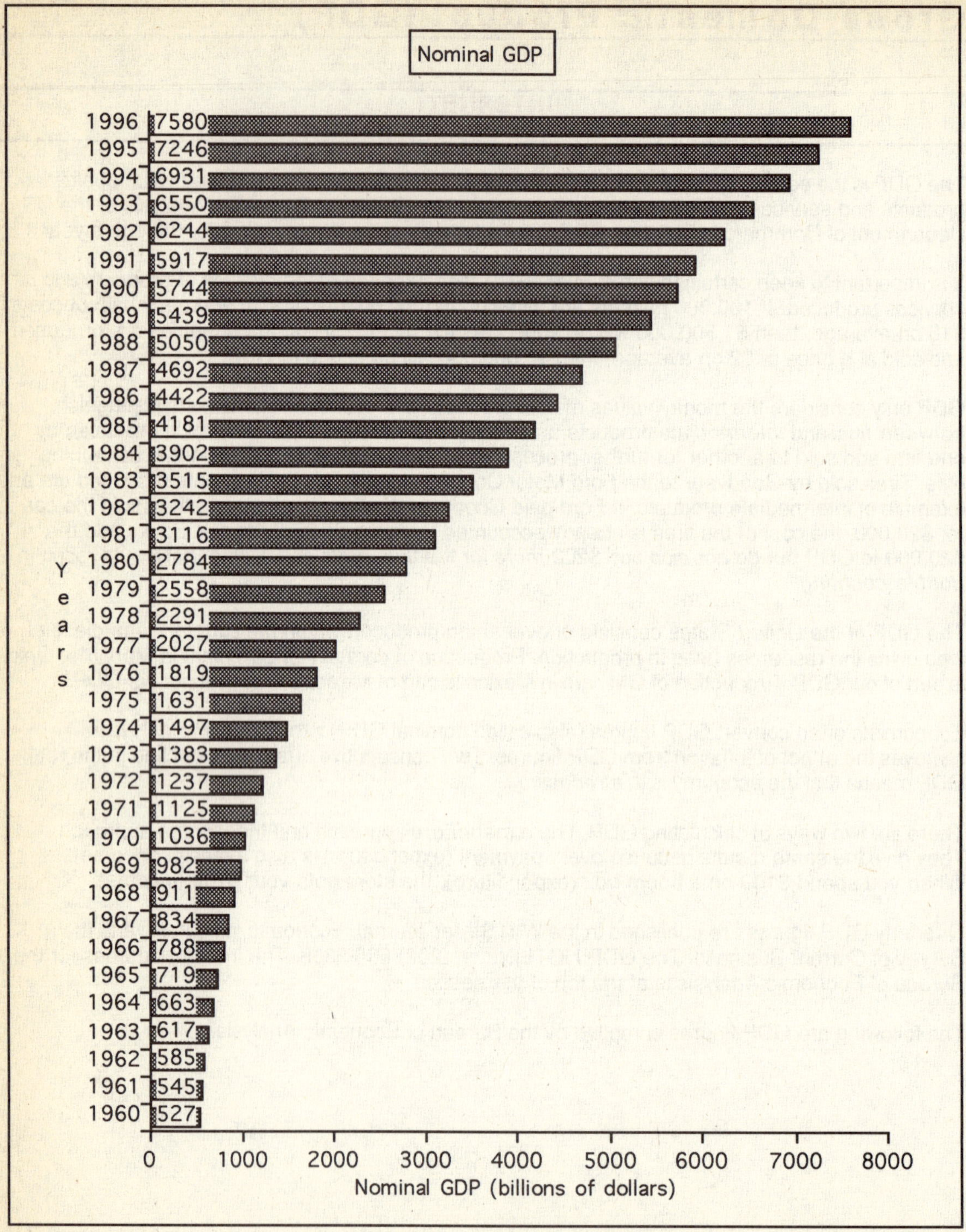
Nominal GDP
1996 7580
1995 7246
1994 6931
1993 6550
1992 6244
1991 5917
1990 5744
1989 5439
1988 5050
1987 4692
1986 4422
1985 4181
1984 3902
1983 3515
1982 3242
1981 3116
1980 2784
1979 2558
1978 2291
1977 2027
1976 1819
1975 1631
1974 1497
1973 1383
1972 1237
1971 1125
1970 1036
1969 982
1968 911
1967 834
1966 788
1965 719
1964 663
1963 617
1962 585
1961 545
1960 527
Years
0 1000 2000 3000 4000 5000 6000 7000 8000
Nominal GDP (billions of dollars)

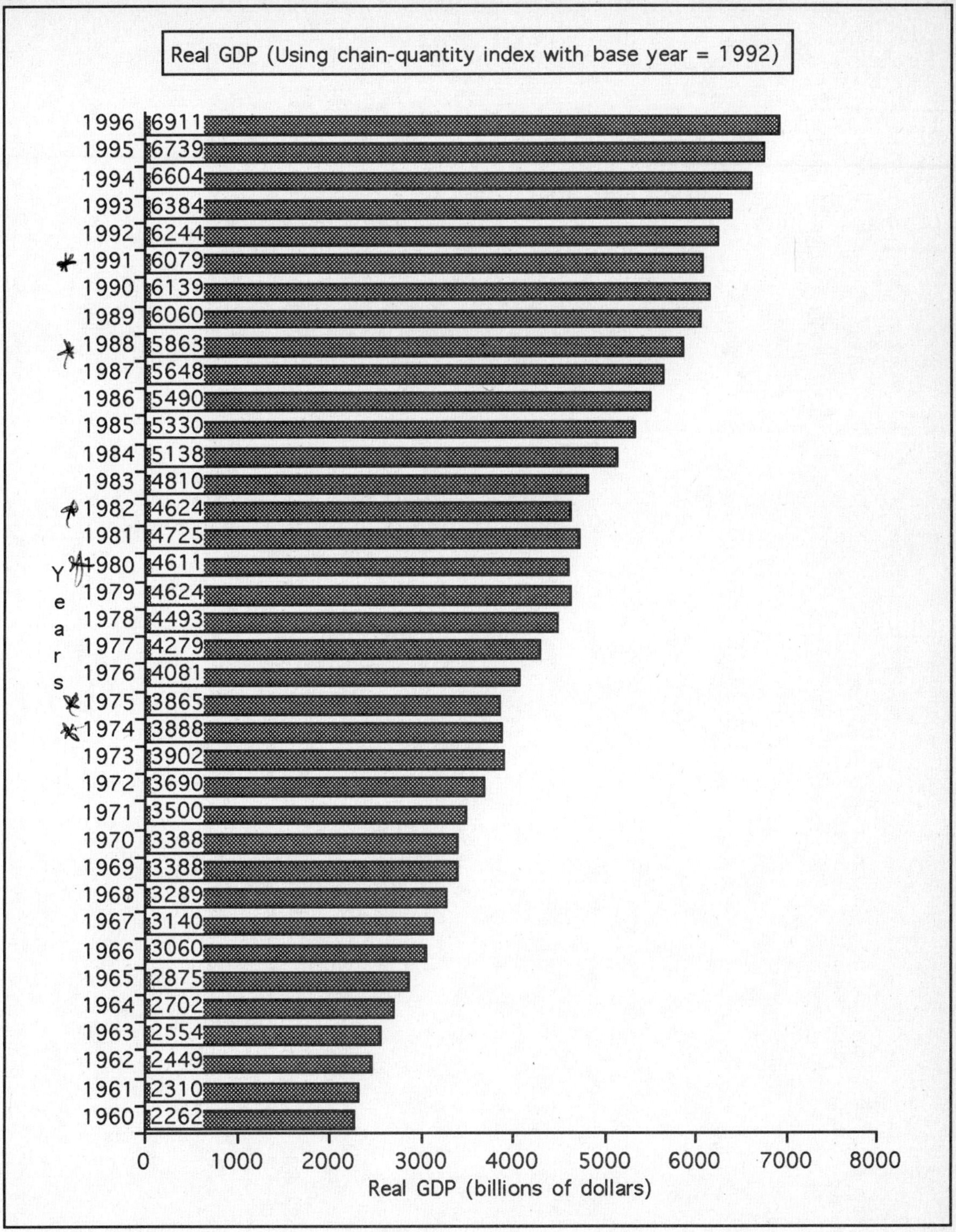
Real GDP (Using chain-quantity index with base year = 1992)
Years
1996 6911
1995 6739
1994 6604
1993 6384
1992 6244
1991 6079
1990 6139
1989 6060
1988 5863
1987 5648
1986 5490
1985 5330
1984 5138
1983 4810
1982 4624
1981 4725
1980 4611
1979 4624
1978 4493
1977 4279
1976 4081
1975 3865
1974 3888
1973 3902
1972 3690
1971 3500
1970 3388
1969 3388
1968 3289
1967 3140
1966 3060
1965 2875
1964 2702
1963 2554
1962 2449
1961 2310
1960 2262
0
1000
2000
3000
4000
5000
6000
7000
8000
Real GDP (billions of dollars)

Unemployment Rate

INTERNET!
http//stats.bls.gov

The unemployment rate for the country is published by the Bureau of Labor Statistics every month. Each month about 60,000 households are interviewed about their recent work history. Workers from the Bureau of the Census equipped with a portable computer conduct the survey which divides the population, 16 years and older, into 4 groups

1) Employed:
People who work for pay or people who work for a family-owned business for profit. This group includes people absent from work because of illness, strikes or vacations.

2) Unemployed:
People not working but either
a) waiting to return to their job from which they were laid off, or
b) actively looking for work for the last 4 weeks

3) Not in the labor force:
These are adults who are not working but not looking for a job. People who are retired, in school full-time or taking care of their home and their children are not in the labor force.

4) Labor force:
The sum of the unemployed and the employed. It consists of people actively participating in the labor market by either working or actively looking for work (or waiting to return to their old jobs.)

The most important statistic compiled from these figures is the Unemployment Rate.

$$\textbf{Unemployment rate} = \frac{\textbf{number unemployed}}{\textbf{labor force}} \times 100$$

Changes in the unemployment rate are very significant because every 1% point *increase* in the rate means that 1.5 million *more* people are jobless. The unemployment figure may underestimate the true extent of unemployment because it is difficult to count *discouraged workers*. These are people who would like to work but have given up looking for work after a prolonged and futile job hunt. Since they are not actively looking for work, they are counted as being 'not in the labor force,' are hence not counted as unemployed.

International comparisons of the unemployment rate are tricky to make. In the US, the government makes a big effort to go looking for unemployed people to count in the unemployment rate. In many European countries, no such effort is made. People have to register themselves with the local unemployment office to get counted as unemployed.

The unemployment rate is compiled by the Bureau of Labor Statistics every month. It appears in the Wall Street Journal, Economic Indicators, Employment and Earnings, Survey of Current Business and several other newspapers and publications. The Unemployment Rate HOTLINE is 202-606-7828. The Internet address for the Bureau Of Labor Statistics is at the beginning of this topic.

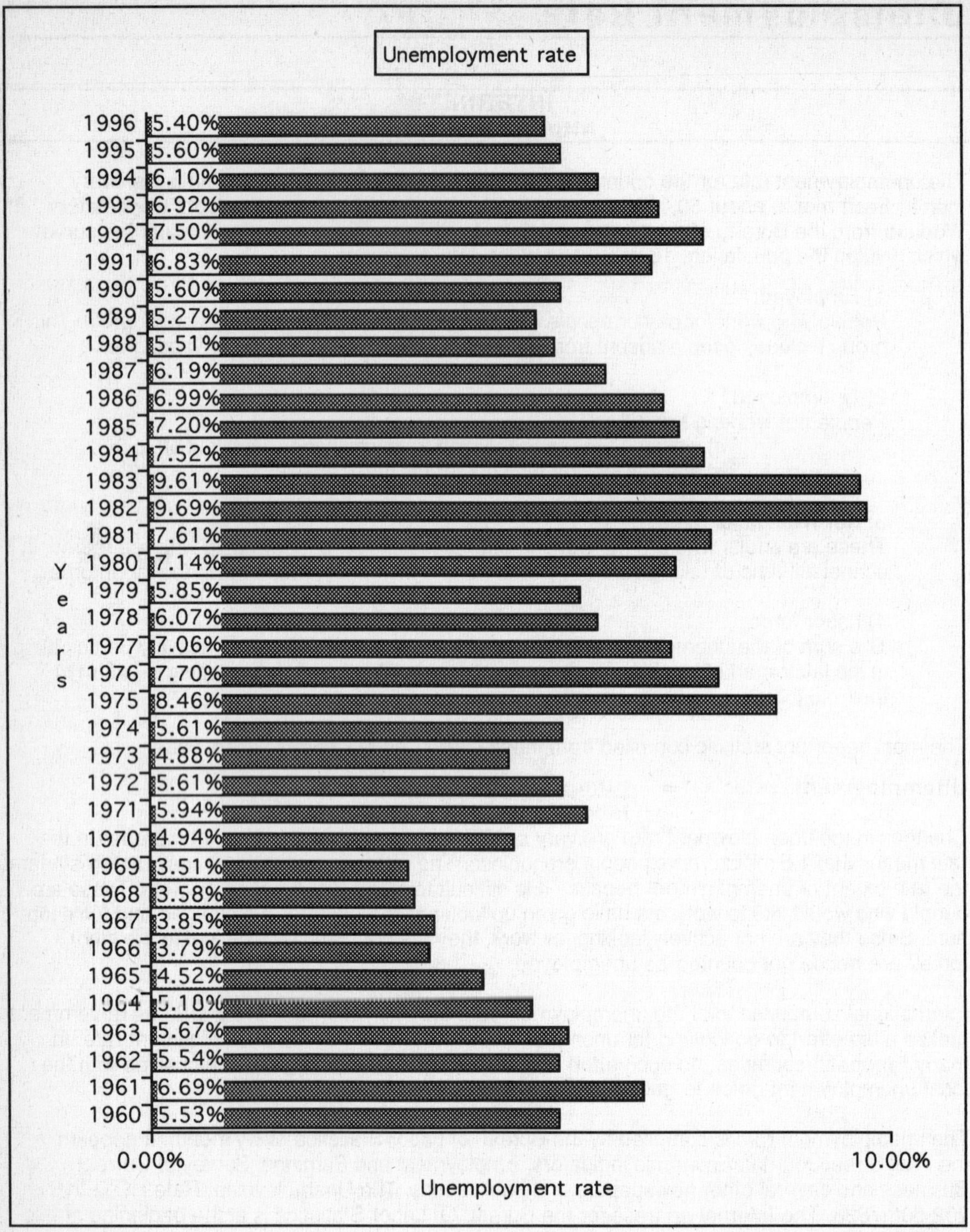
Unemployment rate
1996 5.40%
1995 5.60%
1994 6.10%
1993 6.92%
1992 7.50%
1991 6.83%
1990 5.60%
1989 5.27%
1988 5.51%
1987 6.19%
1986 6.99%
1985 7.20%
1984 7.52%
1983 9.61%
1982 9.69%
1981 7.61%
1980 7.14%
1979 5.85%
1978 6.07%
1977 7.06%
1976 7.70%
1975 8.46%
1974 5.61%
1973 4.88%
1972 5.61%
1971 5.94%
1970 4.94%
1969 3.51%
1968 3.58%
1967 3.85%
1966 3.79%
1965 4.52%
1964 5.10%
1963 5.67%
1962 5.54%
1961 6.69%
1960 5.53%
Years
0.00%
10.00%
Unemployment rate

Consumer Price Index (CPI)

INTERNET! http//stats.bls.gov

The consumer price index (CPI) shows the average level of prices in any period compared to a base period or base year. The prices here are the prices of about 90,000 individual products and services that the typical urban consumer buys.

The Bureau of Labor Statistics actually puts out two different measures of the consumer price index. The first one is the CPI for all urban consumers (CPI-U). When the CPI is reported in newspapers and on television, it is the CPI-U that they report. The second measure is CPI for urban wage earners and clerical workers (CPI-W).

The index is based on a 'basket' or collection of products and services that consumers buy. Employees of the Bureau of Labor Statistics record the prices of these products and services every month by visiting 20,000 retail stores. All these price changes are then averaged using the method of *weighted averaging.* This means that all the prices are not simply averaged. Instead, products that consumers spend a lot of their income on each month, like housing or food get a greater *weight* in the calculation of CPI. When the price of bread, milk and child-care increases a lot, the CPI increases a lot as well. However, when the price of a Rolls Royce or a private jet increases, the CPI is not affected as much because most of us do not spend a lot of our monthly income on buying these products.

The base period or base year is the reference point or point of comparison. Currently, the Bureau of Labor Statistics uses the prices during period 1982-1984 as 'base period' prices. The CPI for that period is set at 100 (The value of any index during the base period is always = 100). The CPI in all other years are expressed as a percent of the prices in the base period. For example, the CPI for 1994 was 148.2. This means that prices on average were 148.2 percent of the prices in the base period. Something costing $100 in the base period would, on average, cost $148.2 in 1994.

The *inflation rate* is calculated as the percentage change in the CPI. This makes the CPI is a very widely used index. It is used to remove the effect of inflation or *deflate* series such as retail sales and hourly earnings or *index* them. Many people have their earnings tied to the CPI. An increase in CPI leads to an increase in their earnings which is enough to compensate them for the higher prices they face. Nearly 70 million people have their income linked to the CPI in the United States. The CPI has also used been used to adjust income tax brackets since 1985.

The CPI figures are made available every month by the Bureau of Labor Statistics. They are published in the Wall Street Journal & other newspapers, Economic Indicators, CPI Summary News Release, Dept. of Labor Bulletin Board, the Economic Bulletin Board. The CPI HOTLINES are (202) 606-6994 for a recorded message and (202) 606-7000 to talk to someone from the Bureau of Labor Statistics. The Internet address for the bureau of Labor Statistics is at the top of this section.

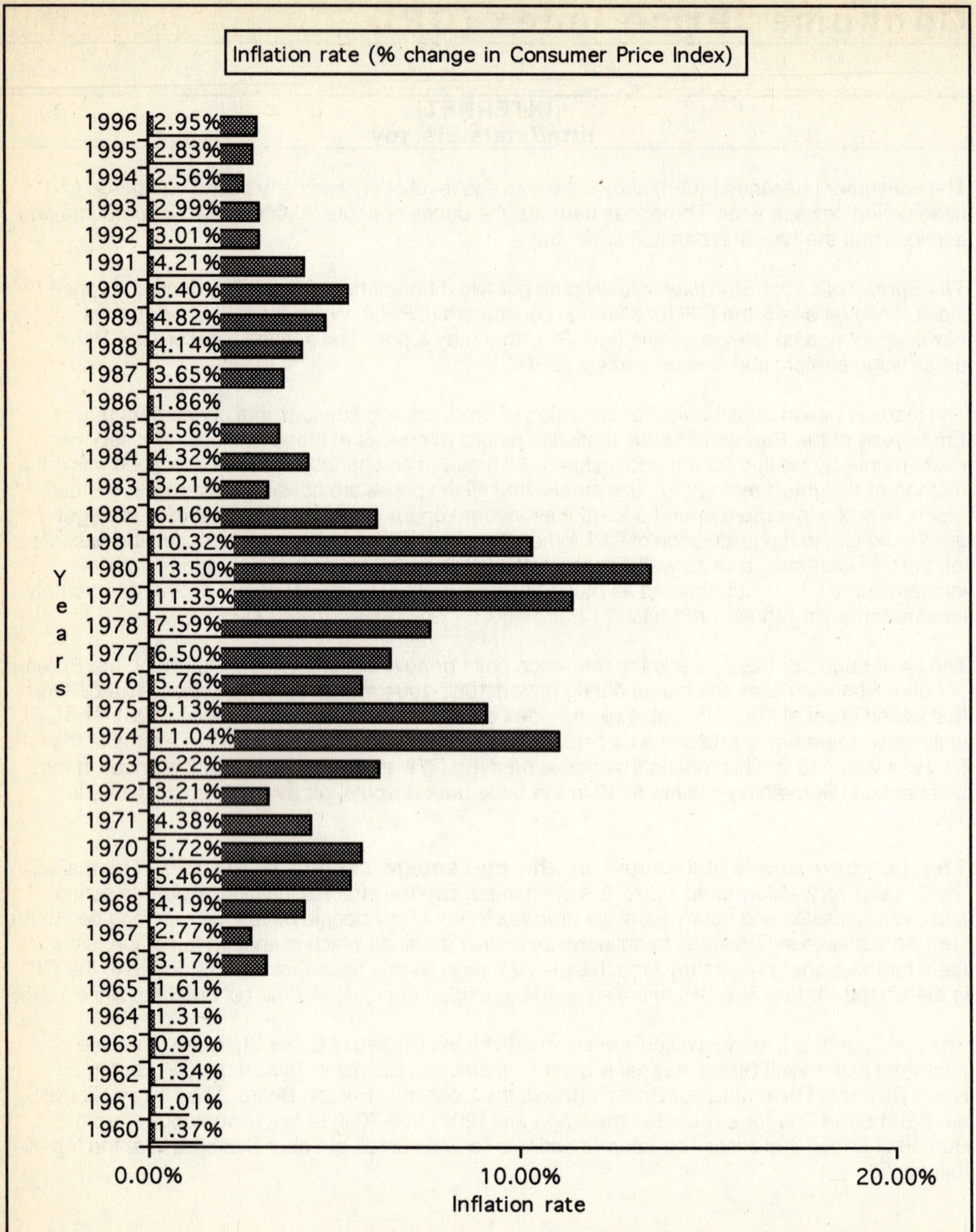
Inflation rate (% change in Consumer Price Index)
1996 2.95%
1995 2.83%
1994 2.56%
1993 2.99%
1992 3.01%
1991 4.21%
1990 5.40%
1989 4.82%
1988 4.14%
1987 3.65%
1986 1.86%
1985 3.56%
1984 4.32%
1983 3.21%
1982 6.16%
1981 10.32%
1980 13.50%
1979 11.35%
1978 7.59%
1977 6.50%
1976 5.76%
1975 9.13%
1974 11.04%
1973 6.22%
1972 3.21%
1971 4.38%
1970 5.72%
1969 5.46%
1968 4.19%
1967 2.77%
1966 3.17%
1965 1.61%
1964 1.31%
1963 0.99%
1962 1.34%
1961 1.01%
1960 1.37%
Years
0.00%
10.00%
20.00%
Inflation rate

Money Supply

INTERNET!
http//www.ny.frb.org/

The money supply is the total amount of money in the economy. When most people think of money, they think of notes and coins, or currency. These constitute only a small proportion of money as measured by the Federal Reserve.

Two of the most important measures of the nation's supply of money are M1 and M2. M1 consists of those items which are actually used to buy and sell products and services.

• M1 includes coins and currency held outside banks, demand deposits, traveler's checks and other checkable deposits.

• M2 includes all of M1, plus savings accounts and money market mutual fund balances.

Of the two, M1 represents the function of money as a medium of exchange most closely. In recent years, however, many of the components of M2 have become more liquid. You can walk up to an ATM and withdraw or transfer funds from a saving account, instantly. So economists now watch both M1 and M2.

Some studies have shown a close link between the level of M2 and the price level, so some economists watch changes in M2 as a leading indicator of inflation.

The money supply figures are compiled by the Federal Reserve Board of Governors every week. They appear in the Wall Street Journal, Statistical Release H. 6 and Federal Reserve Bulletin both published by the Federal Reserve Board of Governors, Survey of Current Business, and several other newspapers and publications. The Internet address for the Federal Reserve Bank of New York is at the top of this section.

The following pages contain data on both M1 and M2.

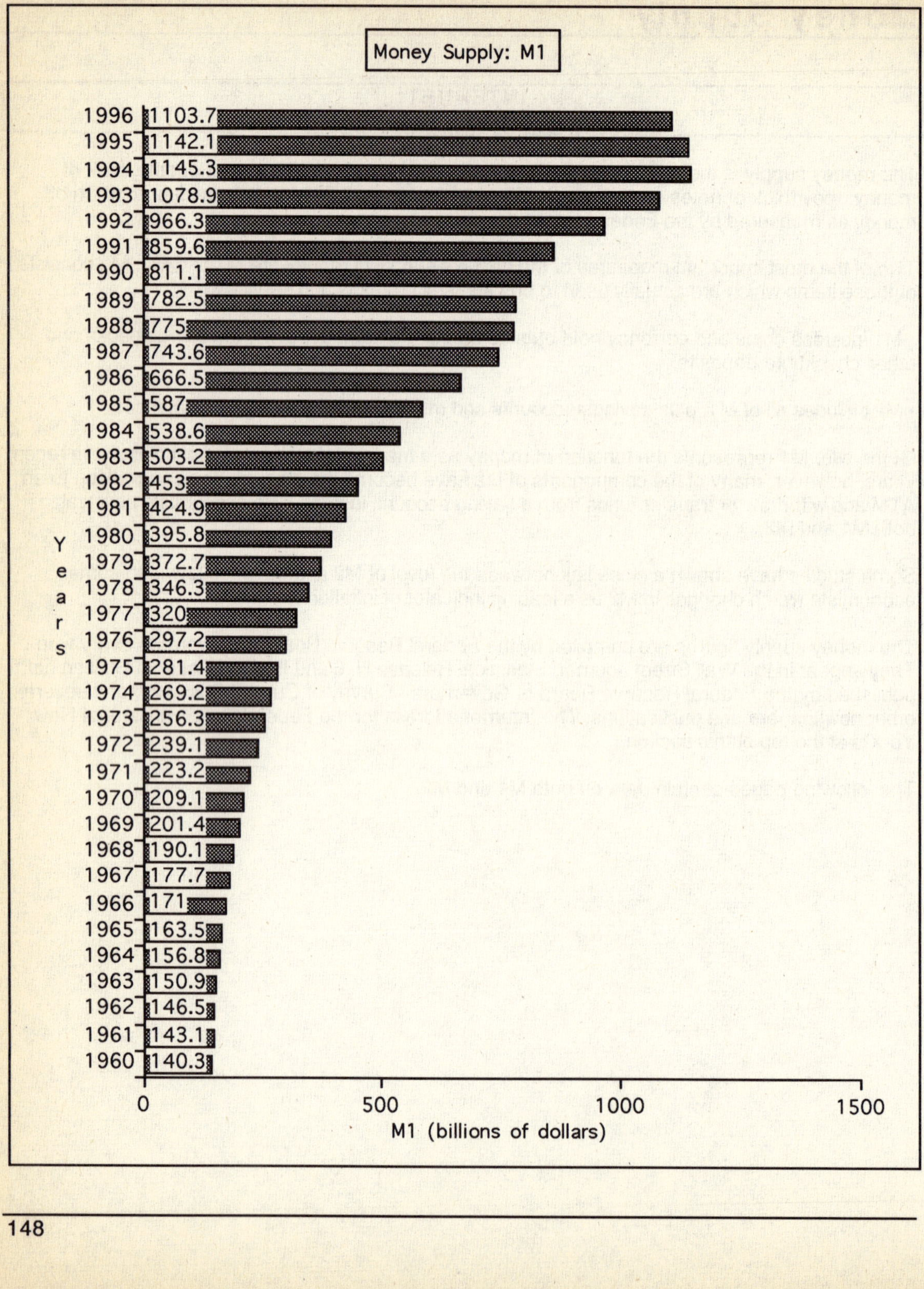
Money Supply: M1
1996 1103.7
1995 1142.1
1994 1145.3
1993 1078.9
1992 966.3
1991 859.6
1990 811.1
1989 782.5
1988 775
1987 743.6
1986 666.5
1985 587
1984 538.6
1983 503.2
1982 453
1981 424.9
1980 395.8
1979 372.7
1978 346.3
1977 320
1976 297.2
1975 281.4
1974 269.2
1973 256.3
1972 239.1
1971 223.2
1970 209.1
1969 201.4
1968 190.1
1967 177.7
1966 171
1965 163.5
1964 156.8
1963 150.9
1962 146.5
1961 143.1
1960 140.3
Years
0
500
1000
1500
M1 (billions of dollars)

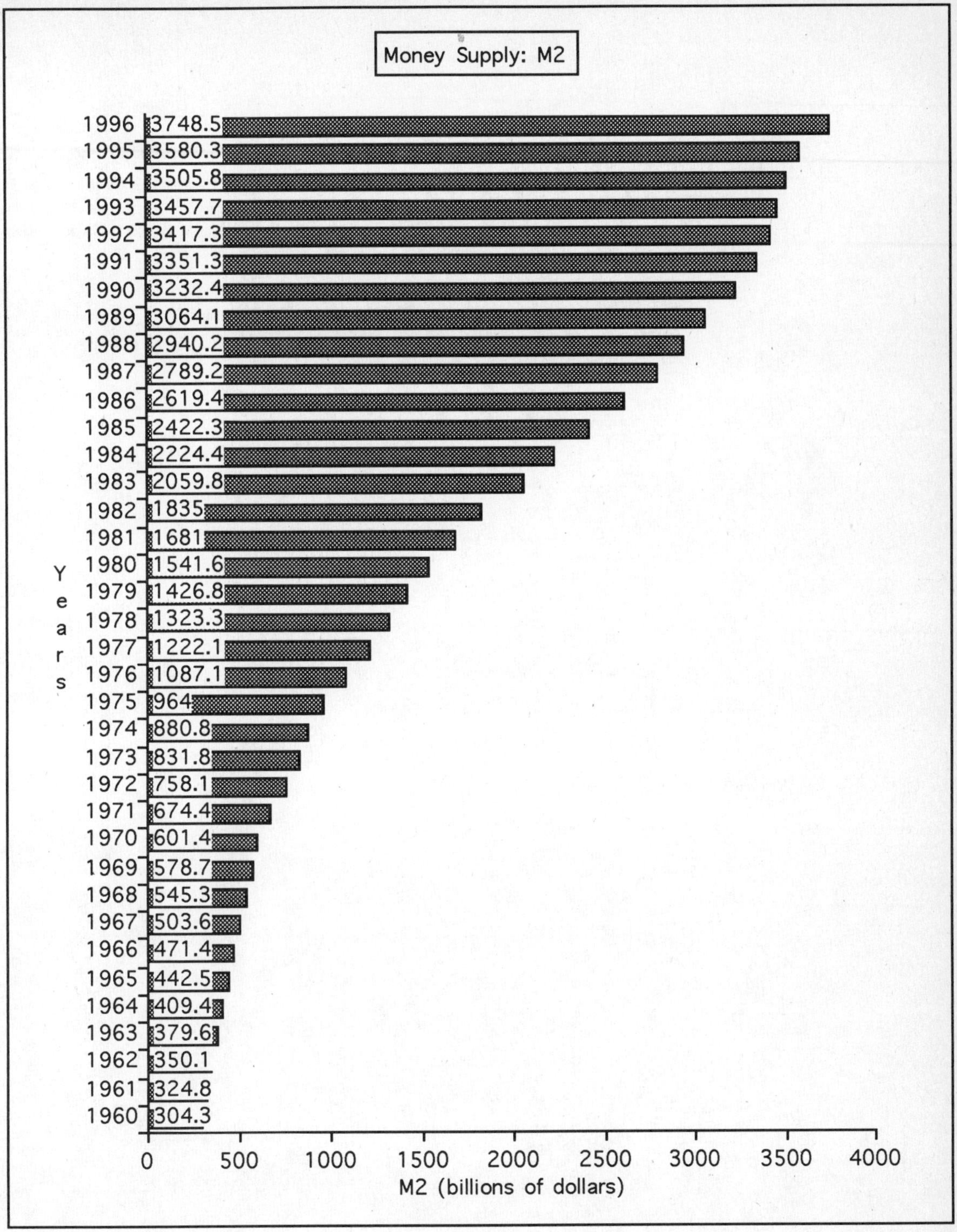
Money Supply: M2
1996 3748.5
1995 3580.3
1994 3505.8
1993 3457.7
1992 3417.3
1991 3351.3
1990 3232.4
1989 3064.1
1988 2940.2
1987 2789.2
1986 2619.4
1985 2422.3
1984 2224.4
1983 2059.8
1982 1835
1981 1681
1980 1541.6
1979 1426.8
1978 1323.3
1977 1222.1
1976 1087.1
1975 964
1974 880.8
1973 831.8
1972 758.1
1971 674.4
1970 601.4
1969 578.7
1968 545.3
1967 503.6
1966 471.4
1965 442.5
1964 409.4
1963 379.6
1962 350.1
1961 324.8
1960 304.3
Years
0 500 1000 1500 2000 2500 3000 3500 4000
M2 (billions of dollars)

Interest Rate

INTERNET!
http//www.ny.frb.org/

There are many interest rates in the economy.

• The prime rate is the interest rate banks charge their best customers.

• The discount rate is the interest rate the Fed charges banks for loans.

• The Fed funds rate is the interest rate banks charge other banks for overnight loans.

Of all interest rates, the Fed funds rate is the most important. It is the first interest rate the Federal Reserve tries to change with its open market operations. The Fed funds rate is the only interest rate that the Bureau of Economic Analysis has classified as a leading indicator for peaks in overall economic activity. This means that as a recession approaches, the Fed funds rate is the first interest rate to turn down. For all these reasons all other interest rates tend to follow changes in the Fed funds rate. By keeping an eagle-eye on this important interest rate, we can track the movements in interest rates in general.

The Fed funds rate is compiled by the Federal Reserve Board of Governors every week. It appears in the Wall Street Journal, Statistical Release H. 15 and Federal Reserve Bulletin both published by the Federal Reserve Board of Governors, Survey of Current Business, and several other newspapers and publications. The following page contains data on the Fed funds rate. The Internet address of the Federal Reserve Bank of New York is at the top of this section.

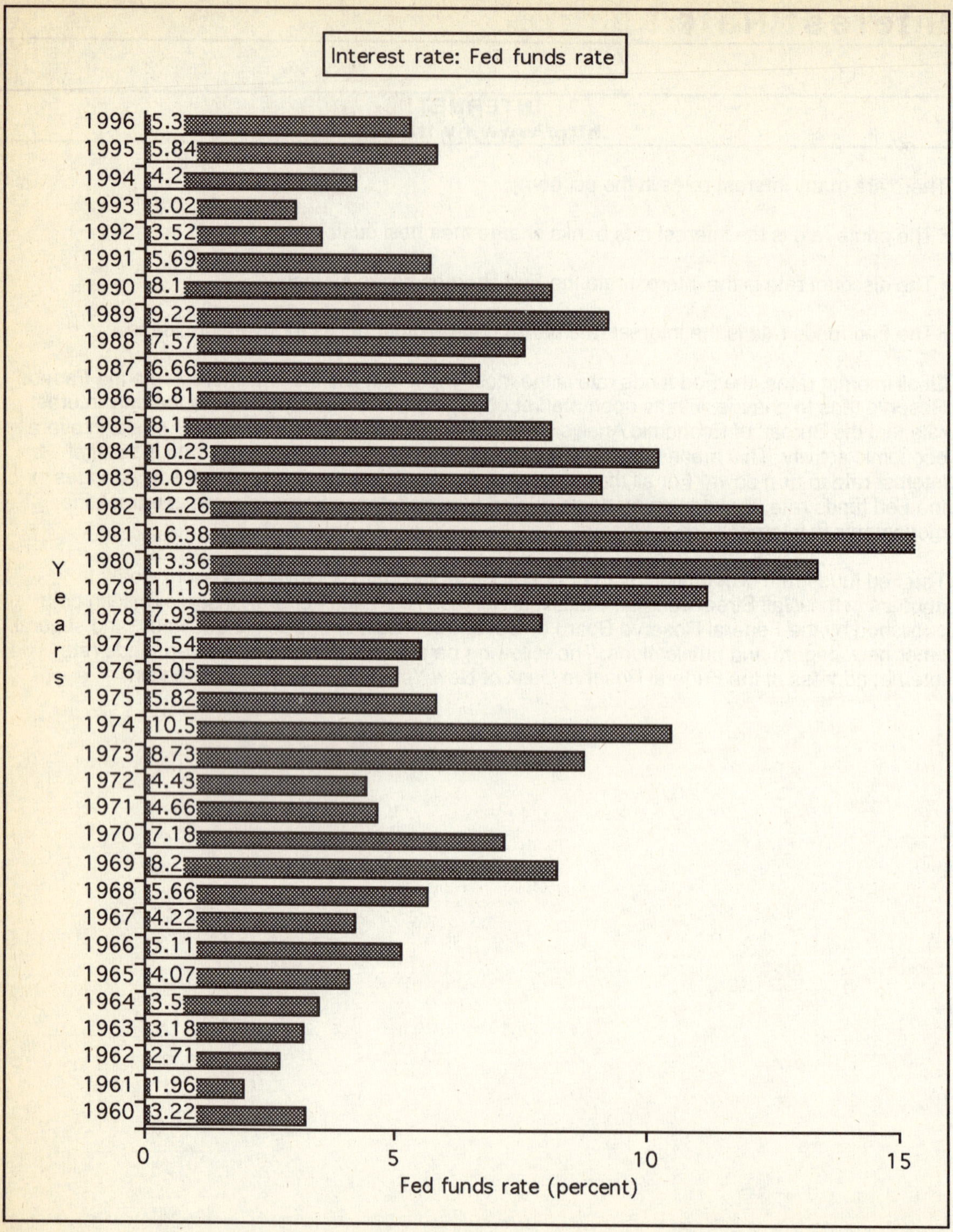
Interest rate: Fed funds rate
1996 5.3
1995 5.84
1994 4.2
1993 3.02
1992 3.52
1991 5.69
1990 8.1
1989 9.22
1988 7.57
1987 6.66
1986 6.81
1985 8.1
1984 10.23
1983 9.09
1982 12.26
1981 16.38
1980 13.36
1979 11.19
1978 7.93
1977 5.54
1976 5.05
1975 5.82
1974 10.5
1973 8.73
1972 4.43
1971 4.66
1970 7.18
1969 8.2
1968 5.66
1967 4.22
1966 5.11
1965 4.07
1964 3.5
1963 3.18
1962 2.71
1961 1.96
1960 3.22
Years
0
5
10
15
Fed funds rate (percent)

Balance Of Trade

INTERNET!
http//www.census.gov/

The balance of trade, or technically, the **balance on merchandise trade** accounts keep track of a country's exports and imports of goods. The word balance seems to indicate some periods of trade deficits and other periods of trade surpluses. For the last twenty years, however, the United States has consistently imported more from abroad than it has exported, leaving the country with consistent trade deficits.

The underlying cause of the continuing balance of trade deficits is the low saving rate in the United States. The low domestic saving rate means that the US must turn to foreigners for its capital needs. A large inflow of capital into the country is then balanced by a large trade deficit.

The merchandise trade balance figures are compiled by the Bureau of the Census every month and quarter. It appears in the Wall Street Journal, Report FT900 published by the Bureau of the Census, Survey of Current Business and several other newspapers and publications. The merchandise trade balance HOTLINES are (301) 763-5140 for a recorded message and (202) 606-5362 for a summary of international transactions. The Internet address of the US Bureau of the Census is at the top of this section.

The next page contains data on the balance of trade.

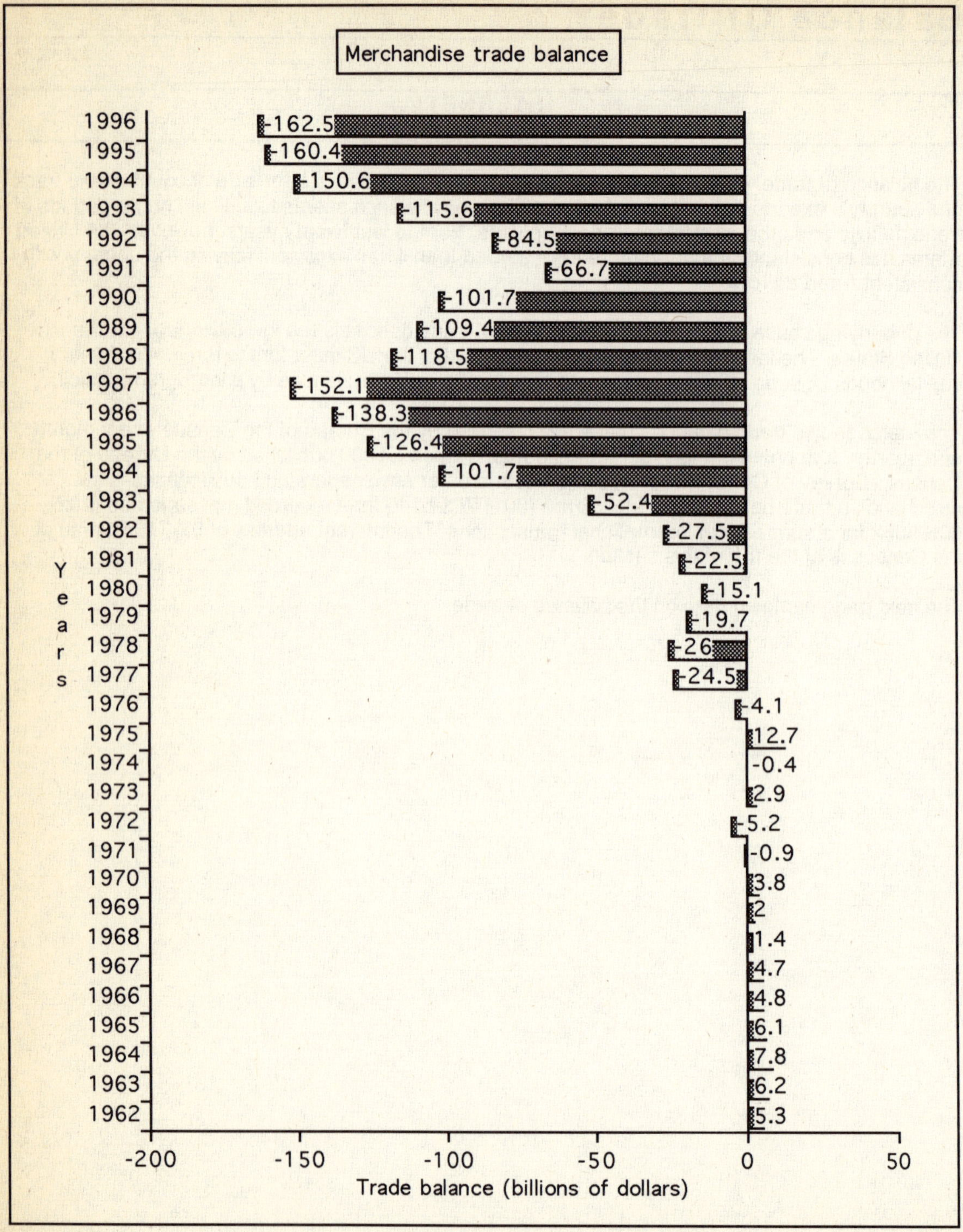

Merchandise trade balance
1996 -162.5
1995 -160.4
1994 -150.6
1993 -115.6
1992 -84.5
1991 -66.7
1990 -101.7
1989 -109.4
1988 -118.5
1987 -152.1
1986 -138.3
1985 -126.4
1984 -101.7
1983 -52.4
1982 -27.5
1981 -22.5
1980 -15.1
1979 -19.7
1978 -26
1977 -24.5
1976 -4.1
1975 12.7
1974 -0.4
1973 2.9
1972 -5.2
1971 -0.9
1970 3.8
1969 2
1968 1.4
1967 4.7
1966 4.8
1965 6.1
1964 7.8
1963 6.2
1962 5.3
Years
-200
-150
-100
-50
0
50
Trade balance (billions of dollars)

SECTION FOUR

Interviews with Economists

Interview with Milton Friedman

Best of Both Worlds
conducted by: Brian Doherty
Reason Magazine

Milton Friedman reminisces about his career as an economist and his lifetime "avocation" as a spokesman for freedom.

Milton Friedman needs little introduction. His career as one of the world's preeminent economists and advocates of freedom has won him many accolades, best-selling books, and a Nobel Prize.

It has also brought him much satisfaction. Now, in what he is acutely conscious are probably the last years of his life, he and his wife and longtime writing partner Rose Friedman are working on their memoirs.

I met Friedman in January in his elegant high-rise San Francisco condo, with an absorbing view of both the Pacific Ocean and the San Francisco Bay. His study is filled, but not cluttered, with his own books and economics reference works. While some Great Men in his position in life might refuse nuisances like interviewers entirely, Friedman is friendly and mostly forthcoming, speaking with the slow assurance of a lifelong professor and teacher very comfortable with explaining things. He welcomed me cordially but with a distinct set of limits, both in time and in subject matter. He has a large project to finish, and not much time to finish it in; and he refuses to psychoanalyze himself, largely avoids indulging in discussion of personalities, and wants to save some stories for his memoirs.

Friedman is used to discussing policy, but except for his assessment of the new Congress's potential, we wandered far afield into reminiscence; assessment of his intellectual development; and his thoughts on the history, significance, and successes of the intellectual movement for freedom that he has served so staunchly.

Reason: You've long advocated many of the ideas the new Congress is pushing, such as balanced budget amendments and flat taxes. Do you think Congress will make your dreams come true?

Milton Friedman: I'm skeptical. The talk is good. But I expected so much out of the Reagan administration and was disappointed. I'm a great admirer of Ronald Reagan himself, and I suspect he would have gotten much more done if it hadn't been for the Cold War and the problem of Nicaragua and El Salvador. But nonetheless, there's no doubt that while he talked about cutting down the size of government, he did not succeed. He did slow it down - you've got to give him credit for some achievements. But not the massive reduction that he hoped for and planned for. That makes me hesitant now.

Congress wants to talk in this direction. Would they really want to move in that direction? The most important reform would be term limits, six-year limits. Because from an economic point of view, one of the worst features of our system is that you have a new tax law every year or every two years. However bad the tax law is, if you didn't change it for five years it would do less harm. Why do you keep changing it? Because that's the most effective way to raise campaign funds. Lobbyists will pay you to put loopholes in; they will pay you to take them out.

If you can get a flat tax with no exemptions or deductions - the Armey plan I suppose would be fine - its main advantage would not be the greater equity of a flat tax or less interference in private incentives. It would be to end this business of changing the whole tax system every few years and keeping prosperous these hordes of tax lawyers.

Reason: You were involved in the development of the withholding tax when you were doing tax work for the government in 1941-43.

Friedman: I was an employee at the Treasury Department. We were in a wartime situation. How do you raise the enormous amount of taxes you need for wartime? We were all in favor of cutting inflation. I wasn't as sophisticated about how to do it then as I would be now, but there's no doubt that one of the ways to avoid inflation was to finance as large a fraction of current spending with tax money as possible.

In World War I, a very small fraction of the total war expenditure was financed by taxes, so we had a doubling of prices during the war and after the war. At the outbreak of World War II, the Treasury was determined not to make the same mistake again.

You could not do that during wartime or peacetime without withholding. And so people at the Treasury tax research department, where I was working, investigated various methods of withholding. I was one of the small technical group that worked on developing it.

One of the major opponents of the idea was the IRS. Because every organization knows that the only way you can do anything is the way they've always been doing it. This was something new, and they kept telling us how impossible it was. It was a very interesting and very challenging intellectual task. I played a significant role, no question about it, in introducing withholding. I think it's a great mistake for peacetime, but in 1941-43, all of us were concentrating on the war.

I have no apologies for it, but I really wish we hadn't found it necessary and I wish there were some way of abolishing withholding now.

Reason: You've also had some history of advising candidates and presidents. How did you get involved in the Goldwater campaign?

Friedman: Through Bill Baroody at the American Enterprise Institute. The American Enterprise Institute was originally the American Enterprise Association, and had established a board of academic advisers to advise them on their publications. I had been a member of that I think since its inception, and Baroody arranged sometime in the early '60s a number of dinners at his house at which Goldwater was present. Baroody was the brain trust for Goldwater. I was also at some of those dinners, so I got to meet Goldwater. And then when the campaign came along, Baroody asked me to serve as economic adviser. I didn't go on the campaign trail. I sat at home and wrote memos.

Reason: Were you impressed with Goldwater's acumen?

Friedman: It depends on what you mean by acumen. There's no doubt whatsoever that he's a man of principle and strong character. His IQ is perfectly reasonable but it's not outstanding among the various politicians I've met, and that shows why IQ is not a good measure. The highest IQ was Richard Nixon's and he was a terrible president.

While I was never a governmental official, I was a member of an economic advisory group that Nixon appointed of which Arthur Burns was chairman. I saw Nixon from time to time when he was president, until he imposed price controls. I saw him only once after that.

Reason: Did you stop giving him advice?

Friedman: I kept giving him advice from Newsweek, but not personally.

Reason: Do you have a clear memory of how your political philosophy formed? Was it any specific teacher you encountered, book you read, or experience?

Friedman: I'm sure it was a combination of all of those. I was exposed as an undergraduate at Rutgers to two very strong influences: Homer Jones, who was a student of Frank Knight's from Chicago, and Arthur Burns. They both had a considerable influence on me as an undergraduate in my thinking and my writing.

But it would be hard to say what philosophy that left me with. One of the things I regretted all my life is that when I graduated from Rutgers and came home, I wrote out a statement of my beliefs. I put that away in a drawer somewhere in my mother's home and I've never been able to find the damn thing! I'd love to have it! So I can't really tell you what I believed at that time.

But obviously my ideas were not very well formed. I was an innocent youngster and what I was impressed by, of course, was the Great Depression, and the belief that somehow or another there ought to be something that can prevent any such thing from happening.

Thanks to Homer, I was offered a scholarship at the University of Chicago and I went to Chicago and studied with Frank Knight, Jacob Viner, Henry Schultz, and so on. The atmosphere in Chicago in 1932 was very lively and active and encouraging. Of course, I got a very good grounding in economic theory and statistics as well.

Next year, I managed to get a fellowship to Columbia. I spent a year at Columbia mainly to study with Harold Hotelling, who was a mathematical economist and statistician.

Then I went back to Chicago for one year and was a research assistant to Henry Schultz. There were a group of students in Chicago who were very, very important. George Stigler, Allen Wallis, Rose Director, and myself. We ate almost every lunch and dinner together. We spent all the time discussing economics, both economic theory and economic policy. And we were very close for the rest of our lives. George died about two years ago. Allen, I'm glad to say, is still alive.

In the 1930s, both Rose and I at separate times went to Washington and worked on the New Deal, but we were technical statisticians and economists, not anything that had any policy role.

Throughout my career, I spent most of my time on technical economics. This policy stuff has been a strict avocation. If you really want to engage in policy activity, don't make that your vocation. Make it your avocation. Get a job. Get a secure base of income. Otherwise, you're going to get corrupted and destroyed. How are you going to get support? You're only going to get support from people who are ideologically motivated. And you're not going to be as free as you think you're going to be.

One of the most important things in my career is that I always had a major vocation which was not policy. I don't regard what I've done in the field of monetary policy as on the same level as what I've done about trying to get rid of the draft or legalizing drugs. One is a technical byproduct of scientific work, and so that's the only sense in which my vocation has affected my policy. But by having a good firm position in the academic world, I was perfectly free to be my own person in the world of policy. I didn't have to worry about losing my job. I didn't have to worry about being persecuted.

I think you'll make a mistake if you're going to spend your life as a policy wonk. I've seen some of my students who have done this. And some of them are fine, and some of them, especially those who have gone to Washington and stayed, are not.

Reason: How did you come to enter the world of policy writing?

Friedman: What really got me started in policy and what led to Capitalism and Freedom was, in an indirect way, the Mont Pelerin Society. The first Mont Pelerin Society meeting was in 1947 in Switzerland. Hayek arranged it. It was his idea.

Mont Pelerin was the first time that I came into contact with people like Hayek, Lionel Robbins, and the European contingent of that time. That widened my perspective about issues and policy.

The Mont Pelerin Society was people who were deeply concerned about issues. It was people with whom you shared a basic common belief, who at home were isolated. Its great contribution was that it provided a week when people like that could get together and open their hearts and minds and not have to worry about whether somebody was going to stick a knife in their back - especially for people in countries where they were isolated.

The reason the Society ever happened was that Hayek had written The Road to Serfdom, which attracted the attention of the Volker Foundation, and it was the Volker Foundation that financed the American participation in the Mont Pelerin Society. A Swiss group financed the Swiss and European participation.

In the middle '50s, the Volker Foundation undertook a program of summer institutes for junior academics who were favorably inclined toward a free-market point of view or were interested in such issues. Capitalism and Freedom was based on a series of lectures that I gave at one of those seminars. Those seminars forced me to systematize my thoughts and present them in a coherent way. And they also provided a very good audience because the people who were there were lively, outspoken, didn't hesitate to criticize. It was a very good audience. There was a lot of free time as well for discussions outside of the formal seminar. And I learned a great deal, not only from the students who were there, but also the fellow lecturers.

And then my wife, Rose, took the transcribed tapes of the lectures and reworked them and that's what became Capitalism and Freedom.

Reason: Did you have any hesitation about publishing that book?

Friedman: None whatsoever. Why should I have had any hesitation? Remember, I was a tenured professor.

Another thing that helped form my policy orientation was when Hayek came to Chicago in 1950. He attracted quite a number of very able students, Sam Peltzman, Ron Hamowy, Ralph Raico, Shirley Letwin. There were quite a group of them. Hayek drew very high quality people. I was an adviser to their New Individualist Review and contributed articles to it. They were a very lively group that had organized discussion sessions and so on, which was part of the atmosphere.

I was persuaded at that time in the early 1960s that we were on the verge of developing a strong libertarian movement. These were libertarians, all of them, though Hayek would not have labeled himself a libertarian. As you know, he always avoided the term conservative, too. He would call himself an Old Whig. The others would have called themselves libertarians.

That's how I was able to develop my own ideas. What shaped them was the interaction with all these other people at lunches and dinners and lectures.

Ayn Rand was receiving increasing attention at that time. I believed a big upsurge in the libertarian philosophy and views was pending. And to some extent it was. You had the Randian group, and the Murray Rothbard group. But the developing libertarian movement was repressed by the Vietnam War and what it led to. You've only got room for one big movement at a time.

Reason: Why do you think you had more initial success as a public proselytizer - you had a regular column in Newsweek - than other prominent libertarians?

Friedman: I really don't know how to answer that. I was basically trained in economic science. I was interested in the history of thought and where it came from. I thought I was going back to some fundamentals rather than creating anything new. Ayn Rand had no use for the past. She was going to invent the world anew. She was an utterly intolerant and dogmatic person who did a great deal of good. But I could never feel comfortable with her. I don't mean with her personally -- I never met her personally. I'm only talking about her writings.

Rothbard was a very different character. I had some contact with Murray early on, but very little contact with him overall. That's primarily because I deliberately kept from getting involved in the Libertarian Party affairs; partly because I always thought Murray, like Rand, was a cult builder, and a dogmatist. Partly because whenever he's had the chance he's been nasty to me and my work. I don't mind that but I didn't have to mix with him. And so there is no ideological reason why I kept separate from him, really a personal reason.

Reason: In seeing yourself as harkening back to 19th-century liberalism, you never became a system-builder like Rand or Rothbard. . . .

Friedman: Exactly. I'd rather use the term liberal than libertarian.

Reason: I see you occasionally use the word libertarian.

Friedman: Oh, I do.

Reason: As a concession to accepted usage?

Friedman: That's right. Because now liberal is so misinterpreted. So I am a Republican with a capital "R" and a libertarian with a small "l." I have a party membership as a Republican, not because they have any principles, but because that's the way I am the most useful and have most influence. My philosophy is clearly libertarian.

However, libertarian is not a self-defining term. There are many varieties of libertarians. There's a zero-government libertarian, an anarchist. There's a limited-government libertarianism. They share a lot in terms of their fundamental values. If you trace them to their ultimate roots, they are different. It doesn't matter in practice, because we both want to work in the same direction.

I would like to be a zero-government libertarian.

Reason: Why aren't you?

Friedman: Because I don't think it's a feasible social structure. I look over history, and outside of perhaps Iceland, where else can you find any historical examples of that kind of a system developing?

Reason: One could argue the same thing about minimal-state libertarianism: that historically it seems to not be stable.

Friedman: I agree. I wrote an article once arguing that a free society is an unstable equilibrium. Fundamentally, I'm of the opinion that it is. Though we want to try to keep that unstable equilibrium as long as we can! The United States from 1780 to 1929 is not a bad example of a limited-government libertarianism that lasted for a long time.

Reason: Is feeling like part of a larger movement important to you? Would you have been able to do the work you did had you not felt part of a community of like-minded scholars?

Friedman: I've been very fortunate in being part of two communities of scholars: the community of economists on the one hand, and the community of libertarians on the other. And that combination has been very productive so far as I'm concerned, but I can't really tell you why. One thing is that it's very hard for somebody on his own to be sure that he's thought of all the angles. Discussion among people helps an enormous amount. And particularly able, good people.

If you have a person isolated in an environment unfriendly to his ideas and thoughts, he tends to turn bitter and self-directed. But the same person with three or four other people around -- it doesn't have to be a lot of people -- will be in a wholly different position since he will receive support from the others.

You remind me of one incident where in a sense the two worlds interacted. Back in the 1960s, my daughter was an undergraduate at Bryn Mawr, and I was invited by Haverford, I think it was, to spend three days giving talks on mathematical economics. Absolutely no policy involved, pure mathematical economics. And because my daughter was at Bryn Mawr, I agreed.

After I had agreed, they asked if I would also be willing to give a chapel talk on political matters. I said sure and I gave a title, something having to do with freedom. Then I discovered that chapel at Haverford was compulsory. I wrote to the president and said that I was very much disturbed at giving a talk on freedom to a compulsory audience.

When it was time to go to the chapel, I asked the president, "How do they count attendance?" And he said, "At the beginning of the hour there are people going around in the balcony and looking down. Everybody has an assigned seat, and they count."

When I got up to talk, I spoke up to the people in the balcony and said that those who were counting attendance, please let me know when they're through because I don't like the idea of speaking about freedom to a compulsory audience. I'm going to sit down and give the people who want to leave the chance to leave. And I did. Now, the students hadn't really thought that I was going to do it and when I did, about one or two people got up to leave and the rest of them booed them because obviously, I was talking on their level. As a result, I've seldom had a student audience who were so completely on my side as that group, even though the political atmosphere at Haverford was very much to the left. That's one of the greatest coups I've ever had as a public speaker.

Reason: Do you think you've become more radically libertarian in your political views over the years?

Friedman: The difference between me and people like Murray Rothbard is that, though I want to know what my ideal is, I think I also have to be willing to discuss changes that are less than ideal so long as they point me in that direction. So while I'd like to abolish the Fed, I've written many pages on how the Fed, if it does exist, should be run.

Murray used to berate me for my stand on education vouchers. I would like to see the government out of the education business entirely. In that area, I have become more extreme, not because of any change of philosophy, but because of a change in my knowledge of the factual situation and history.

I used to argue that I could justify compulsory schooling on the ground of external effects. But then I discovered from work that E.G. West and others did, that before compulsory schooling something over 90 percent of people got schooled. The big distinction you have to make is between marginal benefit and average benefit. The marginal benefit from having 91 percent of people in school rather than 90 percent does not justify making it compulsory. But if in the absence of compulsory education, only 50 percent would be literate, then I can regard it as appropriate.

Some issues are open and shut. Tariffs, property rights. No, not property rights, because you have to define property rights. But education is not open and shut. In Capitalism and Freedom we came out on the side of favoring compulsory schooling and in Free To Choose we came out against it. So I have become more radical in that sense. Murray used to call me a statist because I was willing to have government money involved. But I see the voucher as a step in moving away from a government system to a private system. Now maybe I'm wrong, maybe it wouldn't have that effect, but that's the reason I favor it.

Reason: Would you agree with the proposition that you have been the most successful and important proselytizer for libertarianism?

Friedman: I don't think that I've had the most influence. I think the most influential person was Hayek. The effect of The Road to Serfdom was really critical. In another area, Bill Buckley has certainly been very important on national policy.

Buckley's not a libertarian. But he's also not a socialist. And if you look at the political scene, his National Review has had a tremendous influence in providing a base for collaboration between the libertarians on the one side and the free-market conservatives on the other. That was epitomized in its most obvious form by Frank Meyer when he was with National Review. They've helped that coalition to form and hold together and have influence; Bill Buckley played an enormously important role.

I might have more public influence than ideologues like Rand or Murray Rothbard, the libertarians in that strict sense. And I believe that the reason is because they have been so intolerant.

Reason: You wrote an essay in Liberty about the intolerance of Rand and Ludwig von Mises. You say you never met Rand. . . .

Friedman: I was never to my knowledge in the same place as she was; I was in Chicago, she was in New York. I'm sure if I had been in New York, I would have met her. It was not because of any objection on my part. I think she was a fascinating woman and had a great influence. As I always have said, she had an extremely good influence on all those who did not become Randians. But if they became Randians, they were hopeless.

Reason: But you knew Mises personally. Did you see the intolerance that you find in his method also in his personal behavior?

Friedman: No question. The story I remember best happened at the initial Mont Pelerin meeting when he got up and said, "You're all a bunch of socialists." We were discussing the distribution of income, and whether you should have progressive income taxes. Some of the people there were expressing the view that there could be a justification for it.

Another occasion which is equally telling: Fritz Machlup was a student of Mises's, one of his most faithful disciples. At one of the Mont Pelerin meetings, Fritz gave a talk in which I think he questioned the idea of a gold standard; he came out in favor of floating exchange rates. Mises was so mad he wouldn't speak to him for three years. Some people had to come around and bring them together again. It's hard to understand; you can get some understanding of it by taking into account how people like Mises were persecuted in their lives.

Reason: You don't link yourself openly to certain aspects of the libertarian political movement. . . .

Friedman: Well, you have to be more specific. Being very specific, I have not wanted to join the Libertarian Party simply because I have accumulated good working relationships with people in the Republican Party, and I think I can be more effective by being a Republican. That's the only reason. There are no other cases in which I have had any problem with the libertarian movement.

Reason: You certainly have a respectability and presence that most people and organizations labeled libertarian don't have. . . .

Friedman: That's because of one thing only: I won the Nobel Prize. What, are you kidding yourself?

Reason: Your status preceded your winning the Nobel.

Friedman: I did have some of it, yes. It's because I have a firm root in something other than ideology. Because I was firmly based in a scientific academic discipline. I wasn't simply a preacher or an ideologue or an unconnected philosopher.

But I think the libertarian movement is doing fine. I think that *REASON* magazine has been remarkably good; it has been very effective. It takes many kinds of people to make a movement. And one of the most important things are publications. In any activity you have manufacturers, wholesalers, and retailers; and all three are essential and necessary. There are only a relatively small number of manufacturers of ideas. But there can be a very large number of wholesalers and retailers.

As I look around me I'm impressed by the fact that there's increasing attention paid to libertarian ideas. If you look at the picture now, compared with 30 years ago, there's no comparison. Now you've got much more. As far as journals are concerned, then we had the Foundation for Economic Education's Freeman; for a while we had the New Individualist Review in Chicago, but that was about it. Bill Buckley established National Review, which is in a different corner.

But look at the situation today. You have *REASON* magazine, you have *Liberty* magazine. You've got all of this stuff that spouts out from the Cato Institute and the Competitive Enterprise Institute and a half dozen other think tanks. In fact, I think there are too damn many think tanks now.

Reason: Why do you say there are too many?

Friedman: You don't have the talent for it.

Reason: Do you consider yourself in the libertarian mainstream on foreign policy issues?

Friedman: don't believe that the libertarian philosophy dictates a foreign policy. In particular I don't think you can derive isolationism from libertarianism. I'm anti-interventionist, but I'm not an isolationist. I don't believe we ought to go without armaments. I'm sure we spend more money on armaments than we need to; that's a different question.

I don't believe that you can derive from libertarian views the notion that a nation has to bare itself to the outside without defense, or that a strong volunteer force would arise and defend the nation.

Reason: What did you think about the Gulf War?

Friedman: I always had misgivings about the Gulf War, but I never came to a firm decision. It was more nearly justified than other recent foreign interventions, and yet I was persuaded that the major argument used to support it was fallacious.

After all, if Iraq took over the oil, it would have to do something with it. If they don't want to eat it, they'd have to sell it. I don't think the price of oil would have been much affected. The more important consideration was the balance of power with Iran and Iraq. I have mixed feelings about that war; I wouldn't be willing to write a brief on either side.

Reason: What would you regard as your most important accomplishment?

Friedman: It depends on what you mean. I wrote an essay on methodology in 1953. It was published in my book Essays on Positive Economics. I had been working on it for years before that, so it goes way back to the middle '40s. It started to generate a lot of comments, but I decided I would rather do economics than talk about how economics are done. So I made a distinct point of not replying to any criticism of that essay. And I think that's why it's so commented on.

That methodology article has probably been reprinted more often and referred to more often than anything else I've written, though I would by no means regard it as the most important thing I've ever done.

In terms of sheer technical quality there's no doubt in my mind that the best thing I ever did was The Theory of the Consumption Function which, from a scientific point of view, is a carry on from the methodology article. I regard the theory of the consumption function as a demonstration of applying the methodology I explained there. But also it has a neatness about it and a specific theorem which has generated an enormous amount of work since then. When things like that originally come out, the status quo says, "Oh, that's a bunch of nonsense, we can't possibly work with that," but give it time. And by now it's part of conventional economics.

In the realm of policy, I regard eliminating the draft as my most important accomplishment.

Reason: Have you retired from economics?

Friedman: Well, not from economics, but from that kind of work. There's been a tremendous advance in specialization in economics, particularly in the econometrics area. I was just looking at recent working papers published by the Federal Reserve Bank of Chicago. These are clearly built on work of mine, going back to the 1970s. But there's been a new development in econometrics that I haven't kept up with. The techniques they've adopted here are all different from ours. I'm not an expert in them anymore; I really couldn't deal with this material on the level on which they are dealing with it, although I can understand the thrust of what they're doing.

I'm not making any pretense of trying to do any more basic, fundamental economics work. I believe that almost all important contributions of a scientist are made in the first 10 years after he enters the discipline. Not the first 10 years of his professional life; he may shift from one discipline to another. And I've been impressed as I've been going over my memoirs, that my basic contributions all have their roots in the early years of my work. I was reading over some preliminary professional papers in the 1950s, and I could see there the whole future of the next 30 years of work that I did; it was all outlined in there.

You add things to it, you change it, but the fundamental ideas come early. The 1940s-'60s was when I did my most important economic work, even though it wasn't all published then.

Reason: I read an article recently in the Washington Monthly that repeated all the silly ideas about inflation that you've been fighting your whole career. Are battles like this ever won?

Friedman: No. All battles are perpetual. You go back in the literature of economics, and you'll find the same kind of silly statements 100 years ago, 200 years ago. And you'll find the same sensible statements the other way.

Reason: Are those kind of mistakes still made among professional economists?

Friedman: If you look at the views of the profession as a whole, no. There's a great deal of agreement among economists, contrary to what people may think. You won't find much difference of opinion on the proposition that raising the minimum wage will cost jobs. You won't find much difference of opinion on the desirability of free trade. And you won't find any difference of opinion on the idea that you cannot have inflation without monetary expansion. There's no doubt that there's very widespread agreement about those simple ideas.

Reason: How do you make that consensus spread to the general public?

Friedman: You just have to keep on trying to do it. There's no short cut. There's no way in which you're going to end the discussion, because new generations arise; every group has the same crazy ideas. I get a great many letters from people who think that the way to solve budget problems and fiscal problems is to simply print money and pay off the debt. And there's almost no way of making those people realize just what a bunch of nonsense that is.

I'm inclined to think that there's no field so rife with cranks as currency and money, but I'm sure there are other fields that are just as bad. I'm just ignorant of them.

Interviewed by Brian Doherty. Brian Doherty is assistant editor of REASON.

Interview with John Kenneth Galbraith

Unique Perspective: Economic Trends In Today's World
conducted by Shirin Sinnar and Valerie Jaffee
Harvard International Review

Few economists have enjoyed as long or as distinguished a career as John Kenneth Galbraith, professor of economics at Harvard since 1948. An early Keynesian, Dr. Galbraith entered the public arena as the chief price controller for Franklin D: Roosevelt during World War II. During the 1950s Dr. Galbraith campaigned for Adlai Stevenson and John F. Kennedy, serving as Kennedy's ambassador to India from 1961 to 1963. Later that decade, he reentered the political spotlight as a prominent critic of the Vietnam War. Dr. Galbraith became best known for his controversial liberal economic views, first receiving attention in his 1956 work, The Affluent Society. *Dr. Galbraith continued to challenge neoclassical economic theory with his books* The New Industrial State *and* American Capitalism.

From the beginning, Dr. Galbraith criticized what he saw as the weaknesses of free market capitalism. He saw American capitalism as enriching most of society while leaving the public sector underfunded and the lowest class marginalized. He also rejected the mainstream idea that the market always "cleared," arguing instead that industrial economies often reached equilibrium below full employment. Dr. Galbraith continues to attack the "conventional wisdom," a term he himself coined, in the articles and books he writes today. Special Features Editor Shirin Sinnar and Staff Writer Valerie Jaffee spoke with Dr. Galbraith in January at his home in Cambridge, Massachusetts.

Harvard International Review: In your 1994 work, Journey through Economic Time, you discussed the globalizing influence of international capitalism and the sense of global association it creates. What do events of the last few years say about the momentum of this trend of economic and political globalization?

John Kenneth Galbraith: I see the trend toward closer trade, cultural, artistic, and travel relationships as inevitable and as something we very much want. I even support the multilateral corporation because when you have a company functioning in several countries, there is an incentive for those countries to avoid stirring up animosity. So I see what is called "globalization" as a civilizing force. On the other hand, that process must be reconciled with the important role of the national state in protecting the welfare of people. When internationalization occurs, there is always a tendency for industry to migrate to those countries where the wages are lowest and where the welfare services are cheapest.

Thus, we have a problem. We have a conflict between two objectives, both of them good. On the one hand, we have to accept the internationalization of economic life, and, on the other hand, we must continue to protect employment, unemployment compensation, social security, health care, and the other services provided by the national state. That is why we must, as in the Maastricht treaty, move toward the unification of economic policy as part of the larger global trend.

HIR: As you are well aware, the present international scene is characterized by large gaps between the quality of life-in developed countries and developing countries. How persistent do you think these gaps are? Is economic convergence at all occurring, or are these gaps widening?

Galbraith: I spent a good deal of my life on this problem. I started the first courses on economic development that were taught at Harvard, and I have spent a good part of my life in the less-developed lands. I consider this inequality to be one of the great problems of our time, particularly as it affects Africa. The African countries were given independence without the educational preparation, without the economic discipline, and without the political discipline that is absolutely essential for economic improvement. This has been one of the tragedies of our time. I have seen the problem at closest range in India where the picture has, especially recently, been much better. But, even in India, there were enormous barriers to the fulfillment of the economic and social hopes of the newly independent country, and even after almost fifty years of independence many of those problems are still unsolved.

HIR: What impact do these economic inequalities have on international relations and security?

Galbraith: I do not believe that they have much of an adverse effect on relations between the rich countries and the poor countries. But there is no question that poverty is a stimulant to conflict. People who have easy lives generally get along well together. People who find life miserable are more willing to risk their lives in ethnic, racial, and religious conflict than people who are more fortunate. I have always strongly supported the idea that developed countries should help underdeveloped countries, and I believe that the main line of support should be through education. In this world, as I have often said, you have no literate people who are poor and no illiterate people who are anything but poor.

HIR: You mentioned the economic experience of India, which has recently moved away from a mixed economy and embraced economic liberalization. For some years, you have favored a more moderate course over full liberalization and suggested a mixture of some "socialist" elements with a free market. With that in mind, how do you feel about India's recent program of economic liberalization? What does this change indicate about the best course of action for other developing nations?

Galbraith: I was a good friend of former Indian Prime Minister Jawaharlal Nehru, and I was at one time attracted by his idea of a mixed economy, which we both drew largely from the same source, English Fabian socialism. But it was clear that that ideal created an administrative problem with which the Indian state could not contend. Therefore, the liberalization of the Indian economy, the reduction of regulation, and the welcoming of foreign investment were all changes that I had come to favor strongly by the time I became an ambassador. Once, when I was ambassador, I was visited by Professor Milton Friedman, and I told him I thought that India was one of the few countries in the world, other than the Soviet Union, where his ideas could make a contribution to well-being. Thus, I would like to see that process of liberalization continue. On the whole, the social controls of India, with many exceptions, have been a barrier to economic development. The exceptions have been in the areas of agriculture, in a range of public utilities, and in public education, which have been very important.

HIR: Moving on to the developed world, you have written extensively about economic conditions in the industrialized nations of North America and Western Europe. How would you appraise the current state of these nations' economies, and how do you see them evolving in the future?

Galbraith: I always avoid predictions because people invariably remember your wrong predictions and dismiss your right ones. But there are some economic trends in the United States which I presently deplore. We are having a war against the poor, a war by the comfortable against the unfortunate. We see that in the effort to cut back social services and the effort to cut back taxes on the affluent. You may conclude that I am not in favor of those efforts, and that is the correct conclusion. On the other hand, I think that this is an episodic development, and that we are already beginning to see the social damage and the political cost from the "Republican Revolution." And so I think the prospect ahead is better than the recent past.

This trend is not confined to the United States; we see this to some extent in all of the Western countries. We have seen a strong example of it these last weeks with labor unrest in France. The Canadians are cutting back on the welfare state, particularly in the province of Ontario, and this is, of course, an acute issue in other industrial countries. The deeper fact is that as the economy develops and people's incomes increase, you have more and more people who lead comfortable lives. Thus, instead of the narrow tier of old-fashioned capitalists, one has a large comfortable class defending their rights against the poor. I dealt with this issue in a book called The Culture of Contentment. I'm not advertising my own books, however.

HIR: On a related note, one of the hottest economic topics in Western Europe today is the issue of economic integration of the region. The Maastricht Treaty of 1991 set the stage for more open trade, and the current leaders of France and Germany are aiming for a common currency. What sort of impact does that integration have upon the trends that you have been discussing?

Galbraith: I regard this integration as an inevitable and welcome trend. It takes us away from the cruel world of the first half of the century, when France and Britain and Russia were locked in conflict with Germany, and gives us the alternative of a more peaceful Europe. We still see, as in the former Yugoslavia and to some extent in Eastern Europe, problems of ethnic conflict. But this is far less serious than the national conflict of the earlier half of this century. I am old enough to remember both World War I and World War II, and I see the greater unification of Western Europe as one long step away from those disasters.

As optimistic as I am about the prospects for greater unity, the establishment of a common European currency is distant. There will only be a common currency when the European nations have some common fiscal policies--tax and expenditure and budget policies--and when they have taken some steps to unify their welfare systems. The common currency immediately attracts the attention of people as a social and political goal, but it actually is one of the later stages, rather than one of the first stages, of unification.

HIR: Regarding the economies of the post-communist countries of Eastern Europe and the former Soviet Union, you have criticized the rapidity of "shock therapy" reform. Do you believe that recent political developments in Poland and Russia, with the political ascendance of formerly Communist leaders, indicate that there will be a retreat from shock therapy? Would such a retreat be desirable?

Galbraith: In 1990, the year after the fall of communism in Europe, I went to a meeting in Brussels of European socialists and social democrats. I argued there that the economic transition of the formerly socialist countries should be carefully considered, rather than launched too rapidly, and should have the strong support of the rich countries of the West. I argued that as there would be trauma involved in moving from comprehensive socialism to market capitalism, the transition should take time. I argued that certain industries should remain under state control for a considerable period of time and that privatization should be a practical, results-oriented matter, not something controlled by ideology. I consider it one of the great misfortunes of our time that we thought that there was some magic about the market system that would enable it to quickly solve all problems. And then, unlike the years following World War II, when we had the Marshall Plan, the rich Western countries have not offered a sufficient level of help.

HIR: One heated issue of US foreign policy is the question of how the United States should deal with governments with severe human rights records, such as those of Cuba or China. Some observers say that the United States is morally obligated to isolate such countries by such means as the denial of trade benefits. Others argue that opening up such countries to trade and international contact will encourage prosperity and thus, in the long run, lead to improved human rights and political freedom. What is your stance on that dilemma?

Galbraith: I would lay down no absolute rules; as with so many matters, this is a question of individual situations rather than of anything that can be decided by formula. There are cases where economic pressure of one sort or another is certainly desirable. I thought that was true in the case of South Africa. On the other hand, there are cases where punishment could only have an adverse effect. I would not, for example, deny Most Favored Nation trade status to China. I think that move would be as likely to arouse anger among a rather proud people as to accomplish the result that is sought. If North Korea were to plead for closer economic relations with other countries and attempt to open up its economy, I would be willing to accept that, even though I would not find it possible to applaud its government.

HIR: As someone who was an early and vocal critic of the US involvement in Vietnam, how do you feel about US and NATO policy toward Bosnia at the moment?

Galbraith: I was sent by President Kennedy to Vietnam in 1961, at the very beginning of our involvement. He sent me because he knew that I did not have an open mind--that I had already decided that involvement in that conflict was unwise. I never changed that opinion. I argued for it inside both the Kennedy and the Johnson administrations and later decided that the only viable course of action was to go public with my views. But Vietnam was in some ways a special case, and I strongly support the Clinton administration in its effort to pacify Bosnia. I consider the Dayton, Ohio treaty settlement of November to be a strong step in the right direction. I am now somewhat optimistic about the prospects for peace there and have been made even more optimistic by NATO action, including that of the United States.

HIR: In *The Culture of Contentment*, you described US foreign policy as a "recreational discipline" for an American elite. In the last few years, since the end of the Cold War, has that foreign policy establishment changed?

Galbraith: I meant for that book to be somewhat annoying to a number of people who dealt with foreign policy issues at the Council on Foreign Relations: In the years following World War I, there was a group of people who thought of themselves as having a monopoly on all thought concerning foreign policy. But I think that, on the whole, foreign policy is now in as good hands as it ever was. That old monopoly has been broken, and I am pleased by that fact.

HIR: You have suggested a decline in the economic power of the United States in recent years. What are the implications of that decline for the international position of the United States?

There were some aspects of the exercise of American power that lead me to feel no regret at the sight of decline. I do not regret seeing a decline in the militarization of other countries, which, beginning in the Dulles years, was very expensive and damaging, and still too pervasive. I would like to see militarization decline even further. Certainly, the economic position of the United States in the decades immediately after World War II gave this nation a power that has since diminished, but I do not regret that. Life all over the world would have been worse if we had not had the recovery of Germany and Japan, the increased stability and improvement in Western Europe, and the great economic development in the Pacific Basin. All of these trends have invaded the economic monopoly that the United States had in the late 1940s and 1950s, and I cannot regret that for a moment. The world is better as a result.

John Kenneth Galbraith has had an distinguished career in academia and public. John Kenneth Galbraith is Professor Emeritus in the Economics Department at Harvard University.

Interview with Paul Krugman

Drift, Diversion, and Snake Oil
by A.J. Vogel
Across the Board

Economist Paul Krugman casts a cold eye on government that deals with trivial issues in major ways and major issues in trivial ways.

How satisfying to have the title of your book become part of the language.

Paul Krugman had that satisfaction with his The Age of Diminished Expectations *(The MIT Press). First published in 1990, a new and revised edition of* Diminished Expectations *was issued earlier this year, then followed in short order by* Peddling Prosperity *(W.W. Norton). His latest book is a frontal attack on what Krugman calls the policy entrepreneur: "the economist who tells politicians what they want to hear." These entrepreneurs, he says, range from left (Robert Reich and Lester Thurow) to fight (Arthur Laffer and Jude Wanniski).*

Krugman, 41, a professor of economics at the Massachusetts Institute of Technology (MIT), defines himself as a liberal - "that is, I believe in a society that taxes the well-off and uses the proceeds to help the poor and unlucky" - but he says that the "fault line between serious economic thinking and economic medicine, between the professors and the policy entrepreneurs, is at least as important as the divide between left and fight."

A.J. Vogl, editor of Across the Board, *spoke by telephone to Krugman at his office at Stanford University, where he currently serves as a visiting professor at the Graduate School of Business.*

Across the Board: From reading your recent articles and books, it's clear you don't have much patience with what's going on in Washington, D.C., today. Indeed, you characterize the government's efforts as a diversion.

Paul Krugman: What we're seeing in some respects, now that we're a bit more than a year into the Clinton administration, is a repeat of what happened during the Reagan administration: That is, instead of dealing with a troubled U.S. economy, we've gone off on a tangent, something that sounds appealing, that plays well on the evening news but that doesn't, in fact, make a lot of sense; something that, in fact, diverts energy and creates problems instead of solving them.

If you take a serious look at the nature of what's wrong with America these days, it's homegrown. It's not a problem of competing on a world market; it's not a problem that's created by people in Tokyo or in Europe. But the administration's biggest focus seems to be on taking a tough line on international trade. In this it's obsessed with a non problem - or a trivial problem, which is trade policy and trade disputes with Japan in particular.

AB: In *Peddling Prosperity*, you argue that the common view of international trade. that it's a war with winners and losers - is wrongheaded.

Krugman: Yes. A lot of people seem to think that countries compete with each other in the same way that corporations compete with each other. One of my favorite quotes on this comes from a speech President Clinton made to Silicon Valley workers in which he said America is like a giant corporation competing in the global economy. That's a view that makes a lot of sense to people if they don't think about it too hard.

But it's really very wrong. I like to explain why by comparing the way Coke and Pepsi relate to each other to the way that the United States and Western Europe relate to each other. If you were to look at Coke and Pepsi, you'd say they were rivals. There's no reason for Coke to see a silver lining in success on the part of Pepsi. A negligible fraction of Coke sales are to Pepsi workers, and a negligible fraction of what Coca-Cola workers buy are Pepsi products. So as far as these two companies are concerned, they are selling similar products in a market, and if one of them succeeds it's to an important extent at the price of the other. It is really a kind of zero-sum game.

Now look at the United States and Western Europe, or the European Union as we have to call it now, and ask if we're competitors in the same way. In the first place, we're not dependent on international trade. An economy like the United States exports only about 10 percent of its Gross Domestic Product. If you were to create a similar number for General Motors - What's the ratio of exports to the gross corporate product of GM? - it actually turns out to be more than 250 percent; that has to do with the fact that we measure output by value added. So first, then, the United States or the European Union - or Japan, by the way, which is not that dependent on international trade either - are basically economies that 90 percent produce things for their own use.

Unlike Coke and Pepsi, big national units are each other's main customers and main suppliers as well as each other's main competitors. As I said, if Pepsi does well, that's purely a competitive threat to Coke. That's it: purely adverse impact. If the European Union does well, that does increase the competitive pressure on U.S. products, but it also expands the market for U.S. exports, and it also improves the supply of goods to you as a consumer.

On balance, it's at least as likely to help the United States as to hurt it; in practice, it turns out that the impact of international competition on the living standards in the United States - or, for that matter, Europe or Japan - has been very small, roughly a wash. So the prevailing view of opinion leaders - that our basic problem is lack of international competitiveness - is just flatly not true given the facts.

AB: If it's not true, then how do you account for its pervasiveness, and also its persuasiveness? A lot of supposedly smart people-not all of them politicians, by any means - become downright eloquent speaking about our trade deficit, the importance of establishing a level playing field with Japan, and so on.

Krugman: Put yourself in a position of a CEO of a corporation that does sell on world markets. Now, you've got to keep your eye on the ball that's relevant for your company. If your company has a foreign competitor or there are several foreign competitors in your market, that's what you're going to be paying attention to. That's an effect: It's really true that you face that foreign competition. However, the other effect - that we are both producers and consumers, and you as a consumer buy foreign products - well, that's much more diffuse. That's not your immediate strategic concern. So it's a terrible thing to say, but I think CEOs have got, necessarily and correctly, what you might call a worm's-eye view of international competition. They're down there in the trenches "fighting a foreign rival head to head," to quote Lester Thurow. They don't have any particular reason to back up and look at the relationships between the economies and see they aren't really competitive at all.

AB: All right, if this international-trade contretemps is a diversion, from what is it a diversion?

Krugman: Let me first say there's a certain limited amount of attention that the government can give to issues. If the Cabinet members are spending most of their time in meetings talking about the next moves in our trade war with Japan, they're not going to be talking about how we should fix the healthcare plan so it actually will work.

Now, there are three levels of damage. The first level is just misplaced priorities. Look at the very limited amount of support for technology development. Where's it going? It's almost entirely focused on manufacturing, which is the sector most favored by this administration because it's the sector that tends to compete in international markets. Well, there's nothing wrong with trying to support manufacturing, but the truth is that ours is mostly a service economy. And the place where we have the most disappointing productivity growth is the service sector. For example, we'd do more for our living standards in the United States by working on the office of the future than by figuring out how to use robots in factories.

AB: What's the second level of damage?

If you've got this misguided belief that international competition is what it's all about, then you naturally develop a kind of belligerence about trade policy that can easily spin out of control. Now, we don't know what will have transpired by the time this interview is published, but it's clear to me - and to other people, like Bill Bradley [the Democratic Senator from New Jersey] - that we have adopted a confrontational stance that is completely disproportionate to what's at stake. It's not that we don't have valid complaints against the Japanese, but we have valid complaints against lots of people and countries for lots of things. That doesn't mean we commence hostilities against them. It's very risky playing hardball at the level we're playing, because once you get into a trade war, it's extremely difficult to get out. We've seen that happen historically: Once countries descend into a spiral of retaliation - even if only a year or two later they realize it was all a mistake it can take decades to undo the damage.

The third level is more diffuse but I think very important: An administration that believes in something that isn't true - and anyone who does his homework realizes it isn't tree - will find it hard to make sensible policy, even in areas that don't seem to have anything to do with that belief.

We saw a lot of that in the Reagan administration. There was a kind of "dumbing down" of economic policy in general that came out of the fact that you really had to be a member of the supply-side faithful to be trusted in that administration. At the risk of naming names, I would say that the most interesting example in the Clinton administration has been the astonishing decision to give Ira Magaziner essentially complete control over the development of the health-care plan. Magaziner's prior policy qualifications were entirely his writing on and consulting on strategic trade. And that work was work that serious economists regarded as completely wrongheaded and uninformed. But in the Clinton administration it was regarded as demonstrating his strategic vision. The result is that we essentially wasted a year on health-care reform, an issue that, in terms of its importance to the U.S. economy, is at least in order of magnitude more important than any positive results we can get from our trade negotiations, even if our aggressive trade policy gets us everything we're demanding.

AB: In *Peddling Prosperity*, you criticize both the supply-siders in the Reagan administration and the people whom you characterize as strategic traders in the Clinton administration. Why don't you put these criticisms in context?

Krugman: The context is simply the 20-year-old stagnation of middle-class living standards in the U.S. economy. That's the starting point, and then there have been two popular movements that have proposed explanations for this stagnation-cures that look to people like me like economic snake oil.

In the 1970s, we had a group of people - the supply-siders - who said the reason why the U.S. economy is in trouble, the reason why we're not managing to achieve the kind of growth that we achieved between World War II and about 1970, is big government. The whole problem, they said, is that the tax rate is too high, and if you just cut taxes everything would solve itself; and, in fact, cutting taxes is such a wonderful medicine for the economy that you don't even have to worry about finding spending cuts to pay for the tax cuts. Just cut taxes.

This was a view that was really coming from only a couple of professional economists, but was enthusiastically picked up - first by the editor of The Wall Street Journal and then by a wing of the Republican Party. I think it's safe to say that as late as 1980 most establishment people, even in the Republican Party, had a hard time taking this stuff seriously. When Ronald Reagan announced that he favored the supply-side platform, his then-contender for the Republican nomination, George Bush, called it "voodoo economics."

But it was, in fact, adopted under the Reagan administration: huge tax cuts without any compensating spending cuts. And, in the end, we discovered it didn't do anything for our long-run problems. Real wages for many workers continued to decline, the long-run rate of the growth of the economy remained inadequate, and poverty continued to rise. And we had a new problem that we didn't have before: a $4 trillion federal debt and an intractable budget deficit that has now become a permanent feature of our economy.

AB: At some point, the strategic traders succeeded the supply-siders?

Krugman: Yes, they emerged during the course of the '80s as a kind of liberal counterpart to the supply-siders. They wanted more government, not less, but they also had a very simple explanation of what all our problems were about. This time it wasn't the government: It was foreign competition. And by adopting an aggressive policy on trade, by promoting high-value industries, we could solve all our problems. In the view of the strategic traders, we weren't doing what was necessary to compete in the global marketplace. Like the supply-siders, the strategic-traders group really contains just a handful of economists, but it struck a chord with some politicians. And, again, I think it's fair to say that it wasn't really taken seriously by the establishment.

AB: You have characterized both supply-siders and strategic traders as peddlers of economic snake oil. Yet a lot of people seem to have bought it, bottled one way or another. Is it that we're looking for miracles, or is something more subtle at work?

Krugman: A lot is that we're looking for one big answer, and a lot of it is the fact that no one else is offering any sort of easy or guaranteed answer. Larry Summers, now Under Secretary of the Treasury for International Affairs, who was expected to become the President's chief economic adviser but was passed over for that, had a pretty good way of putting it: Back in 1983, he described the early works of strategic traders like Robert Reich as "economic Laetrile."

AB: Laetrile? Wasn't that the ultimate cure for cancer?

Krugman: Sure, sure, and the point is we still don't have a cure for cancer. We know some things you can do to avoid it. We have some painful things like chemotherapy that sometimes help. But we can't offer any guarantees, and that's why there's always a market for people who offer miracle cures.

In that sense, we have long-term, difficult economic problems for which the experts, the real experts, have at best difficult and partial answers. We can't say, "Take these two pills and everything will be fine." So there's always a temptation to listen to people offering guaranteed solutions.

AB: Well, if you don't have the hubris required to offer guaranteed solutions, you do have the confidence to tell us what our principal problems are. In your book you say there are really only two.

Krugman: One is that our productivity rate, which averaged 2 1/2 percent a year in the postwar years, up until about 1973, has been more like I percent a year since. That doesn't give you very much margin to have rising living standards.

The second problem, which is related to the first, is that we have growing poverty, which is the consequence both of the fact that the overall pie is growing slowly and that the distribution of the pie is getting more and more unequal. People at the bottom have had a quite rapidly falling share of national income, and we're pushing more and more people over the edge into a really appalling state of poverty, given that we are still a very rich country.

We don't have any magic solutions. Productivity is the very hard problem. One of my favorite comments on productivity growth comes from Robert Solow, the Nobel laureate at MIT, who said that in trying to explain differences in productivity performance across countries and across time, we usually end up in a blaze of amateur sociology. There's a lot we don't know about productivity growth. The best you can say about it is that there's a checklist of things we ought to be doing. It makes no sense for us to complain about low growth while continuing to run large budget deficits, so taxes should be raised and wasteful government programs, like farm subsidies, should be cut. We should reform our deeply distorted health-care system. We should make greater use of innovative schemes like taxes on pollution and congestion that exploit the power of market incentives. No one can tell exactly how much better off we'd be if we did the things on this list, but there's certainly no lack of things to do - probably the most important being the elimination of the budget deficit and working on basic education.

With poverty we have many more levers to use if we're serious about it. One thing that's quite clear is that you can help poor people. You can at least alleviate some of the conditions of really impoverished families. For instance, you can provide nutrition and medical programs for pregnant mothers so their children have a better chance of being born healthy. A relatively small amount of money, because the poor are very poor, can make a big difference there. Of course, you need to structure poverty programs so that they don't provide incentives that lock people into poverty.

I think we understand, or I thought we did - God knows what we understand now, given the way political discourse is being dumbed down - that saying, "Well, there are a lot of low-wage workers out there whose wages are really not enough to get them above the poverty line, so let's increase the minimum wage" - is a counterproductive measure. Yes, those people who still are working will be earning more, but you'll also reduce the number of jobs available. That's an example of screwing up the incentives. Instead, for example, you want to provide earned-income tax credits. Beyond that, we can hope that by improving basic education we not only make our population more skilled but make the distribution of skills more equal, although I have my doubts about how much we can really do on that level.

AB: Improving education along the lines you describe - that requires a rather long-term view, doesn't it?

Krugman: All of these things require long-term views. If I had to zero in on our problems even further, I'd say the one that holds the greatest concern for the future of the United States is not simply poor people but poor children - children growing up in poor families, children who are caught up in a culture of poverty. If you say you're going to try and deal with that, then, almost by definition, there's not going to be visible payoffs in Gross National Product for 20 years, not until these kids really are part of the productive labor force.

AB: I suspect that many executives who read what you've Just said will feel a sense of deja vu, of frustration. It seems that we've put so many anti-poverty programs in place, but what have we got to show for it?

Krugman: Fair enough, but let me say two things: First is that while we've had a zillion programs, relatively small amounts of money have actually gone to poor people. So, if you ask how much support does the United States give to people who are earning very little or who are unable to work, by comparison with any other advanced country, we're spending very, very little.

Second is to understand it's a hard problem and try to do a better job of it. Admit that it's one of our two big problems. Keep our focus on it.

AB: Even with a sharp focus, you seem dubious about what government can do. In your book, in response to your rhetorical question - How should the U.S. government set about solving the problems of productivity and poverty?-you reply, "The answer is straightforward: It shouldn't."

Krugman: Sometimes we actually do know enough to be able to say, "Follow my advice and your problem will be solved." We could do that in the case of malaria, for instance. Other things being equal, it'd be better to seek fundamental solutions to our economic problems, but it's no use insisting that economic policy face the big issues when you have no good idea what to do about them. Where I say that in the book I quote Raymond Chandler: Other things being equal, he said, the bigger the subject of a book, the more interesting it is. But there have been some pretty bad books written about God, and some pretty good ones about making a living while staying fairly honest.

AB: The connection between productivity and international competition is of keen and continuing concern to senior executives. In your book, you pose three questions about this relationship, then tell the reader his answers probably will be wrong.

Krugman: The relationship between productivity and competitiveness is very different from a lot of people's gut reaction to it.

The first question: Are you in trouble if your productivity is worse than that of another country? If, for instance, your productivity is half that of the country next door, does that cause problems? Now, it's important to be clear about what we mean by causing problems. If your productivity is half that of your neighboring country, will your standard of living be lower than your neighbor's? Of course it will. But that would be true if you had built the Great Wall of China between the two countries and there was no contact between them at all. Low productivity means a low standard of living. The question is: Does the fact that the neighboring country has higher productivity aggravate the cost of having low productivity in your country? The answer is no.

If you look at the evidence and think it through, you discover that having a highly productive country next door at least offers you some opportunity to alleviate the consequences of your own low productivity - primarily because your productivity isn't going to be equally low across the board. So the availability of international trade means that you can concentrate on doing the things that you do not as badly as others - and import the things you're particularly bad at doing. The classic economist's example would be something like this: What if Janet Reno is actually a better typist than her secretary? Well, it still makes sense for him to do her typing and for her to run the Justice Department. Maybe he doesn't type as well as Reno, but he does it better than he does being an Attorney General. That sounds trivial, but it's the reality: that a low-productivity country would love to he a high-productivity country, and should try to become one if it knows how, but the fact that it's got neighbors that are more productive is actually a plus, not a minus.

AB: The second question: What happens to a country whose productivity growth lags that of its rivals?

Krugman: That's a twist on the same thing, but it raises some different issues. In the United States, by the way, our productivity is still the highest in the world, but our growth rate has lagged behind it and other countries have been catching up. Let's try a thought experiment: Supposing that every nation in the world had 1 percent productivity growth, what would be the trend in our standard of living? The answer is that you'd expect it to rise about 1 percent everywhere. Now, just suppose that the rest of the world's productivity accelerates at 3 percent and we don't match it. A lot of people's instinctive reaction would be that we're in big trouble; after all, a corporation that doesn't keep up with its rivals' productivity gains is not going to stay in business. That suggests our economic health is dependent on our rate of productivity growth relative to other nations. But think it through: To a pretty close approximation, the rate of growth of our living standard equals the rate of growth of our domestic productivity. Full stop. Period. How fast productivity is growing elsewhere, and whether we're ahead or behind the pack, is irrelevant. It's not a competitive game. We're competing against ourselves, not the world.

AB: For your third question you ask what's more important - productivity growth in sectors that must compete with foreigners, or sectors that are basically producing for themselves.

Krugman: Put it this way: Which would you rather have - ad doubling of our productivity in the electronics industry or a doubling of our productivity in the fast-food industry? Again, most people's instinctive reaction is to say what matters most are the industries that are going head-to-head in world markets. But, again, if you think it through carefully, the truth is U.S. living standards would benefit much more from an increase in fast-food productivity than in electronics, for the simple reason that we have many more people employed in fast food than we do in electronics.

AB: But wouldn't our failure to be competitive in electronics contribute to our trade deficit?

Krugman: It's a little hard to believe, because if we think about a particular industry, we may say, "If we get more productive, we're going to be able to roll back imports and export more," but there's no particular connection between the level of productivity and a country's trade balance. To illustrate, look at Mexico. If you go back to the 1980s, when Mexico was in the throes of a debt crisis, you'd find that during those years Mexico was running big trade surpluses because the country's economic prospects looked bleak. Nobody would lend it money. Bankers who had lent it money were demanding payment, and Mexico was being forced into running a trade surplus essentially to pay its bankers. It did that with a very weak peso, a very undervalued Mexican currency, plus a recession that depressed the level of imports.

After about 1990, as the reforms of the Salinas government took hold and people began to be much more optimistic about Mexico, and as some U.S. companies began demonstrating that Mexican workers could achieve high levels of productivity, money began pouring into Mexico, which enabled it to import lots of capital goods, and it began running large trade deficits.

Well, would you really want to say that Mexico was highly competitive in the '80s and became uncompetitive in the '90s? Actually, that trade surplus in the '80s was a sign of weakness and the trade deficit that it has been running in recent years is a sign of strength. It's a vote of confidence by international business.

AB: Let's conclude by looking into the future. You've just published a revised and updated version of your *Age of Diminished Expectations* and I note you still posit three possible scenarios for the 1990s: "hard landing," "happy ending," and "drift." Which is most likely?

Krugman: We could have a hard landing if international investors pull the plug on our imports of capital the way they did on Latin America in the '80s. I guess, given the current context, we should also throw in the possibility of some kind of really damaging impact stemming from the international-trade conflict, although I suspect that, even if it does get quite nasty, its effect will be more insidious than catastrophic. Then there's the possibility of an oil crisis: World consumption is up again, and we're as dependent as ever on supplies from the Middle East.

But let me say that the reason I don't put too much weight on hard-landing scenarios is that the U.S. economy is a very big, very robust entity. It's not easily triggered into crisis. We've seen its resilience in the face of everything from the 1987 stock-market crash to the oil crises in the past. And the kinds of irresponsibility that we're engaging in like the budget deficit - are not really world-class. That is, we're not running Italian-level deficits, we're not running Brazilian-level deficits, we're not running the kinds of deficits that would make the federal government insolvent in the lifetime of most of the readers of your magazine.

AB: So much for a hard landing. On the other hand, we have a happy-ending scenario, the return of a vigorous economy. To some extent the deterioration of U.S. performance after 1970 is a mystery; and, since it's a mystery, who's to say the magic that mysteriously disappeared won't mysteriously come back again?

Krugman: Sounds rather like wishful thinking.

Well, it's not entirely wishful. The most hopeful argument is the one made by Steve Roach, co-director of global economic analysis at Morgan Stanley, which is that we are just about to actually start figuring out what to do with all this information technology. We've got all these local-area networks, we've got all of these PCs on everybody's desks, but we haven't really changed the way our offices do business.

But we're about to start. There's an historical precedent - the use of electrical power in factories, which was eventually a big source of productivity gain. We went from the old mechanical factory with huge arrays of axles and pulleys to the modem factory in which most machines are driven by stand-alone motors. That meant a big improvement in the layout of factories, material flow, and so on, but it took a long time for people to figure out how to redesign the factories to take advantage of the new technology. So you have to believe that, sooner or later, this astonishing information technology we've been developing will yield a huge productivity payoff.

AB: But you say that the chances for such a happy-ending scenario are about 20 percent.

Krugman: Yes, My estimates on that have been fluctuating. We had a very good productivity year in '92 and I thought this was the coming of the millennium. But now I think, well, maybe not yet. Make that 25 percent. The interesting thing is that public policy is not doing anything to make that happy ending more likely. We're counting purely on the dynamism of the private sector and luck.

AB: What should public policy be doing?

Krugman: Working on the big drags on our economy - productivity and poverty. Which leads me to the most likely scenario: that we just don't really come to grips with things. We don't do anything major to control the budget deficit. We have a few tax increases and spending cuts to hold the deficit within bounds to avoid crisis, but don't really move significantly toward eliminating it. And we don't really do anything about education or poverty, except rearrange the rules a little bit. And nothing terrible happens. People may not feel too bad about the way things are going, and there will be no clear political mandate for big change.

AB: Finally, in this drift scenario, what will happen to income? Will the rich get richer and the poor poorer?

Krugman: Yes, and the middle class will tread water, or maybe do a little better. In this scenario, if you were to look at the income numbers, you wouldn't see an economy in crisis, but if you looked at indicators of social distress - like poverty, broken families, crime - they'd be going up over time. I can't think that a nation in which most people see no convincing improvement in the way they live is going to be very good at clinging to the American dream.

Interview with Juliet B. Schor

Are We Really That Lazy?
conducted by Annetta Miller
Newsweek

A Harvard expert says Japan has it wrong.

It's been a tough few weeks for the image of American workers. First, Japanese House Speaker Yoshio Sakurauchi declared that Americans were "lazy." Then, Japan's Prime Minister Kiichi Miyazawa said they "may lack a work ethic." To make matters worse, some American pundits agreed. But a recently published book by a Harvard economics professor argues that, in fact, Americans are industrious to a fault. In "The Overworked American: The Unexpected Decline of Leisure, *author Juliet B. Schor discusses how too much work is affecting everyone from corporate executives to hamburger flippers. She talked with* NEWSWEEK'S *Annetta Miller.*

Miller: Was Mr. Sakurauchi right, Are American workers lazy?

Schor: No. Working hours in the United States have increased substantially over the past 20 years. From the end of the 1960s to the present, Americans have increased the time they spend at work by about 160 hours - or nearly one month-per year. This is equally true for people in glamour jobs and line workers at McDonald's. And it's true for women as well as men.

It's a fallacy that the productivity of American workers has fallen. The level of productivity of the U.S. worker has more than doubled since 1948. In other words, we could now produce our 1948 standard of living in less than half the time it took in that year. That's largely because of technological change, greater investment and increased skills on the part of the work force.

Miller: So what's behind the Japanese officials' statements?

Schor: Their comments were probably made for domestic rather than foreign consumption-possibly to forestall the growing pressure from Japanese laborers to reduce working hours. Japanese employees are asked to work grueling, even inhumane schedules. And right now, the issue of karoshi, death by overwork, is a topic of major concern. As in all industrialized nations, there comes a point when workers start to put a lot of pressure on their employers to reduce working hours. It's happened in this country, it's happened in Western Europe, and now, it's happening in Japan.

Miller: Where did the perception that American workers are lazy come from?

Schor: Employers have been calling workers lazy for decades. America's workers have been blamed for the trade deficit and the country's economic problems. Part of the perception stems from the conservative political rhetoric surrounding government spending and welfare. And, of course, we've heard a lot about lazy workers from the automobile industry. It's interesting that when U.S. politicians said these things there was very little reaction. But now that Japanese politicians say them, suddenly there's a tremendous furor.

Miller: What about the Japanese? Do they work as hard as everyone believes?

Schor: The Japanese do have longer working hours than Americans. The Japanese work something like 225 hours a year more than U.S. workers. But to put it in perspective, it should be noted that U.S. workers put in the second highest number of hours of anybody in the industrialized world. We work about 320 hours more per year than workers in such countries as Germany and France.

One of the things we have to understand about Japanese society is its gender structure, a structure similar to what the United States had in the 1950s. Women do all the domestic labor. Men are in the factories and shops and they're working long hours, but they don't do much work when the come home. They don't take care of children, they don't do housework, cooking, etc. In the U.S., there are both men and women in the work force and both men and women at home. Japanese workers may work longer in the marketplace, but you have to calculate the total amount of work that people do to figure out how hard people are working and how much leisure time they have. We need to look at things like stress and burnout, which are serious problems among U.S. workers.

Miller: Can workers be as productive working shorter days?

Schor: Historically, as the number of hours in the workday has fallen, the pace of work has increased. This is why productivity can rise, even when hours are shorter. When Medtronic, Inc., in Minneapolis decided to give its employees 40 hours' pay for 36 hours of work, it hired no additional personnel but found that output increased. That's something the Kellogg Co. discovered years ago. When company officials made the switch to a six-hour day back in 1930, they were searching for a strategy to cope with the unemployment of the Depression. To their surprise, they found that workers were more productive, on the order of 3 to 4 percent. In some departments, the pace had picked up even more. The workers were pleased, preferring the quicker pace but shorter hours. What companies are saying today is almost the opposite. Employees are being asked to work more hours for the same pay.

Miller: What about critics who say Americans have to work harder now to remain competitive Internationally?

Schor: Competitiveness is mainly a function of productivity per hour. If companies are saying you have to work more hours for the same pay, that's basically saying that wages have to be lower. I think we've seen the fallacy in that. Undermining wages is a problem because it also undermines productivity and undermines demand. If we want to really cure our competitiveness problems, we need to look much more at factors like management, health-care costs and the education system. Hours are sufficiently long for enough workers.

Miller: Whatever happened to the four-day workweek?

Schor: People were talking a lot about the four-day workweek in the 1950s and '60s. But we're not hearing much about the concept anymore. What's happened is that economic performance has worsened the position of workers. We've reacted to our economic troubles with a theory that says you have to lower wages and work harder. That's been a foolhardy path that's undermining our long-term viability.

Where the four-day workweek is becoming a reality, it's coming under the guise of a longer workday, for example, four 10-hour days. Employers are resisting cuts in working hours. They're saying we have to work harder, like the Japanese.

In many countries, employers are obligated to give more holidays than many Americans ever get.

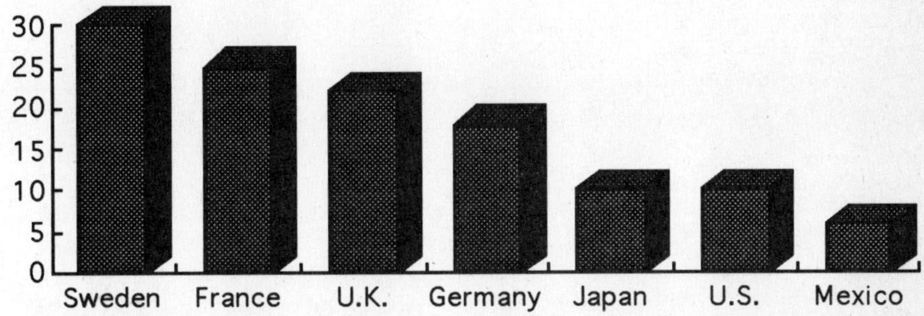

Vacation Time For Employees With One Year Of Service

SWEDEN, 30 DAYS. FRANCE, 25. U.K., 22. GERMANY, 18. JAPAN, 10. U.S., 10. MEXICO, 6.
(Legal minimums except for U.K. And U.S. Source: Hewitt Associates Sotoodeh-NEWSWEEK)

Interview With Lester Thurow

Building a Better Economy
conducted by Sandra Hackman
Technology Review

The renowned maven of industrial policy talks about the threats to the long-term economic health of the United States, and how to counter them by investing in the basic elements of the nation's social system.

With "downsizing" the subject of major media inquiries, including a recent cover story in Newsweek *and a seven-part series of mega-articles in the* New York Times, *economic security is again on everyone's lips. Jobs everywhere seem to be in jeopardy. As economist Lester Thurow points out, reports of continual layoffs only underscore the antiquated nature of our present system. Capitalism, he says, arose when a timely ideology - a stress on maximizing individual welfare - fortuitously merged with the new technology of the steam engine, which could drive an industrial system. Yet today technologies such as computing that rely on human brain power rather than rich natural resources require an ideology that fosters collaboration on crucial social investments such as education and R&D.*

Thurow has never been shy about cutting to the heart of economic issues, whether in revealing the fundamental causes of problems or in prescribing strong medicine for solving them. In magazine columns, in television interviews, and in bestsellers such as The Zero-Sum Society *(1980) and* Head to Head *(1992), he has called for systematic public efforts to guarantee full employment and develop industries on the cutting edge of the high-tech global economy. His latest book,* The Future of Capitalism, *published in March, analyzes the challenges posed by "tectonic forces" sculpting a new economic landscape, such as the infusion of millions of workers from formerly communist countries and the advent of an unprecedented elderly population.*

Formerly dean of MIT's Sloan School of Management and vice-president of the American Economics Association, Thurow is presently Lemelson professor of management and economics in the Sloan School. In an interview with managing editor Sandra Hackman, Thurow talked about the need for the United States to reject self-defeating moves toward "survival-of-the-fittest" capitalism and embrace a thoughtful mixture of public commitment and private initiative essential to creating an economy that works.

Technology Review: Every week we hear of another company laying off thousands of people, yet the unemployment rate is not at its historical high. How can this be?

Lester Thurow: The United States nominally has one of the lowest rates of unemployment in the industrialized world because we don't count a lot of people as unemployed. About 7 and a half million people are officially out of work, but we have another 5 or 6 million people who say they want to work but who don't meet any of the technical criteria for being unemployed, such as that they visited a potential employer in the previous week. We've also got about 4 and a half million people who work part time but want to work full time, and we've got 8 million people who are temporary workers. We have another million people who are self-employed "consultants," many of them professionals affected by, downsizing who may even have one or two customers but who would be glad to take a regular job. We also have about 2 million people who work on call and don't work if their employer doesn't call them. And then we have, between the ages of 20 and 60, 6 million missing men. They're not in school, not at work, not unemployed, not in jail, not retired - they're dropouts.

Overall about a third of the labor force would like to work more than they're now working. If you have that many unemployed, you shouldn't be surprised if the law of supply and demand kicks in and wages drop. Unfortunately, instead of focusing on how to create more and better jobs, the people in Washington are spending all their energy trying to cut the social safety net, arguing that somehow everyone will hold tight and remain in the system. But we know that isn't true - even if some do hang on tight, others will fall off. We also know that people are more likely to succumb to disaster if we take away the safety net, just as when people climb mountains without a rope they are more likely to fall. The fear ratio goes up, and as the fear ratio goes up, coordination goes down.

TR: Do you foresee serious political and social upheavals resulting from significant unemployment and falling wages?

Thurow: Not necessarily. If you reduce wages three-quarters of one percent a year, everybody effectively thinks, "Hey, I'm to blame." It has to do with American ideology. You see this most clearly if you compare this country with France. When the French government recently proposed making what were, by American standards, very small cuts in the social welfare state, the French took to the streets. If you look at the United States, in contrast, Reagan fires all the air traffic controllers and there's no response. AT&T gets rid of 48,000 workers, no response. The guys in Washington make rather large cuts in programs such as Medicaid and student loans, no response. Americans really do believe that if your life goes to hell nobody should come to your rescue, and if somebody does come to your rescue it means you are a lesser human being. We tend to give the system no credit when we do well, so we don't blame the system when we do badly.

TR: Where does this belief come from?

Thurow: It stems partly from the fact that the United States is the only country with founding fathers who we think of as demigods. Europe evolved out of a system of feudalism that everybody understood was wrong and had to change. The interesting thing is how little the U.S. political system has changed in 200 years. For example, everyone knows that American high schools are the worst in the developed world, but we don't do anything about them because Thomas Jefferson invented local school boards and they're deeply embedded in what we think of as a perfect system. Unfortunately, local school boards have conflicts of interest - how do you flunk your neighbors' kids and get elected? - and in any case are uniquely unqualified to run school systems. In general, school boards have no knowledge of what kids are going to need 20 years from now and no interest in the long-run success of the nation. The result is that few Americans could pass the French baccalaureate or the Japanese or German equivalent. Americans who go to college catch up with their European and Japanese counterparts, but those who don't fall permanently behind.

To address this and other long-term problems at the heart of our economic crisis, we need to realize that people's success or failure is due not only to their own initiative but also to how well the social system works. Success is certainly not a function of the "market" alone.

TR: But in your opinion does the United States have to cut back on the social welfare state if we are to create more well-paying jobs?

Thurow: The social welfare state is mostly the elderly, and if you project the number of retired people we will have in 2020, we can't afford to give them what we're giving them today. We're allowing Social Security and Medicare to drive all the investment items out of the budget. So we cut research and development, we cut education, we cut infrastructure, and pretty soon - over a 20-or 30-year period - we have a budget that is all pensions and health care and nothing is left for the investment items that every society needs.

TR: Isn't health care a major economic driver?

Thurow: If you do something for the health care of 20-year-olds you're investing in the economy, because if you can keep them' healthy, they will in fact have long working lifetimes. If you do something for the health care of 70-year-olds, you're not investing in the economy - a 70-year-old's health is pure consumption.

TR: That sounds pretty hard-hearted are you proposing to relegate older Americans to the poverty levels common before Social Security and Medicare ?

Thurow: The issue is not poverty. The average elderly person in America has a per capita income substantially above that of the average person under the age of 65. And of course a lot of people are now retiring earlier. We're subsidizing this income heavily - retired individuals get back all the money they paid into the Social Security system plus interest in less than 4 years. After that they're essentially on welfare. That means we're basically taxing poorer people to give money to richer people.

We should set up a system under which no Social Security recipient gets less than the poverty line but nobody gets more than the median income. If individuals want to do better than that they have to rely on personal savings and private pensions. Unfortunately, politicians don't want to face up to this problem because there are so many elderly voters. The issue is particularly difficult because it is a new one for humanity: the first society in history dominated by elderly voters will occur around the year 2025. But sooner or later we will have to devise an alternative approach because society has to have resources to invest in R&D, infrastructure, and education.

TR: What about cutting the military budget - the other major piece of the pie - to finance pensions and a range of social needs?

Thurow: The military is already down to about 4 percent of GDP. Since Vietnam we've paid for a lot of the expansion of support for the elderly by cutting military expenditures. If you cut those expenditures by another quarter you get a little breathing room but don't solve the fundamental problem.

TR: Why are long-term investments so important in maintaining a successful economy?

Thurow: Take what everybody says is the most exciting economic phenomenon at the moment - the Internet. Where did the Internet come from? It was paid for by the Department of Defense and the National Science Foundation 30 years before it became economically viable. Government basically kept that technology going until it could attract private investment. The same thing happened in biotechnology - the government invested for 30 years before private companies got to the point where they could make profitable biotech products. If we don't make similar investments today, we won't have new industries coming along later that can provide jobs, income, and productivity.

Unfortunately, the budgets Clinton and Gingrich have been fighting over are going to cut investment items proportionately much harder than consumption items and will make our problems worse, not better.

TR: How do we know that if we make those investments they won't pay off more for China, say, than the United States?

Thurow: The problem is you don't. But China is not really competing with Route 128 and other areas that develop high technology. To have a successful economy you've got to have world-class infrastructure, you've got to have world-class skills, and you've got to do world-class R&D, and China simply isn't rich enough to do all those things. So there's room for everyone.

There are some industries - textiles is one of them - where low-wage countries have a tremendous advantage, but there are lots of other industries where low wages don't provide an advantage. The automobile industry is a good example, because in the best automobile plants these days we assemble a car with about 11 hours of labor. If you're building a $30,000 machine with 11 hours of labor, it doesn't make much difference whether you pay the workers $40 an hour or $4 an hour. And you don't do it with labor-intensive methods because you can't build a high-quality car with thousands of people swarming over it.

TR: How do we decide which projects to invest in, as well as which ones government should pursue and which ones should be left to private industry?

Thurow: With R&D there's kind of a natural division of authority. If a project should pay off within 6 years, then that's something private industry can and will finance. If you're talking about something that's not going to pay off in terms of a marketable product within 10 years, it is clearly something government should finance. For between 6 and 10 years there needs to be a mixture. In Europe the government says: if you come through the door with a good medium-term project and half the money, I'll provide the other half. So what they're doing is lengthening time horizons and cutting risks, but the firm has to put up some money.

TR: The Clinton administration hasn't fared very well in its efforts to secure funding for joint public-private R & D projects.

Thurow: They talked about doing that in the beginning and then it all got buried under health care. I think the administration made a big mistake: its three possibilities for major initiatives included a big infrastructure project, reskilling and upgrading the education of those Americans who graduate from high school but don't go to college, and health care.

The administration chose health care. But that was a poor choice. Reforming health care was never something that was going to happen within two years - it would more likely take eight years. Americans don't like paying for health care, but they're also reluctant to change an existing system. Plus health care is a consumption item.

I think the choice should have been infrastructure because it is an investment item, and because projects such as building bridges are visible and can be accomplished within a reasonable period of time. Also, the great thing about infrastructure is that we know how to do it.

TR: What kinds of projects should we fund?

Thurow: One area clearly is high-speed rail transportation between major cities within 200 miles of each other. We could also speed up the process of wiring the country with fiber optics. And we could improve our roads and bridges, which we have basically let rust. Those efforts not only stimulate the economy but they provide a lot of jobs, and they're generally well-paying jobs.

But the most important "infrastructure" investment is education. No private company would ever pay for producing a college graduate. Sixteen years of money in before you get any money out? It doesn't meet capitalist criteria. Yet producing college-educated adults is terribly important in reruns of maintaining a successful economy. And no society in the world has ever become literate without being willing to invest substantial funds in public schools.

Basically we have. to understand that we live in a very different world than we did in the 1950s and '60s - it's now a world that depends much more on knowledge industries than on natural resources, which you either inherit or you don't inherit (and nobody inherited more than we Americans), And those industries, just like the people who work in them, have to be incubated over a long period of time.

TR: Even if we educate Americans better and boost funding for R&D and infrastructure, will enough well-paying jobs result?

Thurow: They will if we allow the economy to expand. The Federal Reserve Board is using high interest rates to hold economic growth to 2.5 percent. To generate good jobs for most people, you have to be willing to forgo the war on inflation that's already been won and let interest rates drop further. You also need a global economy that's growing faster, which means putting pressure on other countries to let their interest rates fall and their economies expand.

TR: Do we also have to use trade negotiations to open up other countries to what we are selling?

Thurow: The problem is not other economies - it's Japan. At some point the United-States and the world have to solve the Japanese problem, which is that nobody has been successful at selling products in Japan. The Japanese have been running a $140 billion trade surplus with the world, about half of that with the United States.

Unfortunately, President Clinton hasn't been willing to threaten the Japanese so they will take seriously his demands that they buy more U.S. goods. What we've been doing thus far doesn't work. Talking doesn't work. Negotiating doesn't work. Yelling and screaming doesn't work.

We have to give up the idea of intervening directly in the Japanese economy while also finding a way to correct today's serious trade imbalance. Rather than trying to force the Japanese to open up specific sectors such as automobiles and rice, we ought to insist simply that they buy roughly as many goods from us as they sell to us. If, for example, the Japanese buy $100 billion worth of our products in any given year, we should allow them to sell $110 billion worth of products to us the following year. That way the ball is in their court - they can decide what they want to import, or even whether they want to import.

TR: What do you think of Patrick Buchanan's suggestion that we impose high tariffs on Japanese goods?

Thurow: I'd rather give the Japanese an incentive to expand their imports than prevent them from continuing to export to our economy.

TR: In any case, shouldn't our first concern be to get our own house in order?

Thurow: Yes. At the moment we're selling the idea of survival-of-the-fittest capitalism while forgetting about the community itself. If you don't make the basic infrastructural and social investments you gradually see a stagnant economy. If you remember your high-school history, teachers used to talk about all these wealthy societies that fell apart because people quit investing in their irrigation canals. The Moors had very elaborate irrigation systems in southern Spain, and the Incas had comparable systems in Peru. When the Spanish conquered those two societies they didn't make the social investment in the canals, and half a century later those once-rich places were very poor.

TR: Are we capable of learning such lessons?

Thurow: That depends. When something very dramatic happens - for example, the 1929 stock market crash societies usually reorganize themselves to prevent it from happening again. What people find much harder to deal with is a very slow decline - say, of one-half of one percent per year - that we look back on 25 years later and say, "Hey, that was a big deal."

TR: How can we snap out of our lethargy?

Thurow: We have to follow Chairman Mao's dictum that a journey of a thousand miles starts with the first step, and the first step is to choose some highly visible project that evokes national pride the way the moon project did - for example, a world class high-speed rail line between San Diego and San Francisco or between Boston and New York. Such projects would not only provide a more efficient transportation system and local jobs but excite people about American technological prowess, underscore the government's commitment to serving its citizens, and awaken us to the possibilities for building a better economic future.